GEOGRAPHERS
Biobibliographical Studies

VOLUME 14

GEOGRAPHERS: BIOBIBLIOGRAPHICAL STUDIES

GEOGRAPHERS
Biobibliographical Studies

VOLUME 14

Edited by
Geoffrey J. Martin

on behalf of the
Working Group on the History of Geographical Thought
of the International Geographical Union and the
International Union of the History and Philosophy of Science

MANSELL

First published 1992 by
Mansell Publishing Limited, *A Cassell Imprint*
Villiers House, 41–47 Strand, London WC2N 5JE, England
387 Park Avenue South, New York, NY 10016–8810, USA

British Library Cataloguing-in-Publication Data
Geographers: biobibliographical studies – Vol. 14
1. Geographers – Biography – Periodicals
910'92'2 G67

ISBN 0-7201-2116-7

Enquiries concerning publications listed in the 'Bibliography and Sources' section of the biobibliographies should be sent to the relevant publisher or journal, and *not* to Mansell.

Typeset by Colset Private Limited, Singapore
Printed and bound in Great Britain by
Biddles Limited, Guildford and King's Lynn

Contents

Introduction

This volume of *Geographers: Biobibliographical Studies* is offered in a new format. It was thought that fewer biobibliographical studies produced at one time would render the whole more digestible to the reader. The essays are designed in the manner to which we have been accustomed since the first number appeared in 1977 under the editorial supervision of T.W. Freeman and P. Pinchemel. Since that time more than two hundred geographers representing more than thirty countries have been rendered in this format. The purpose of this biobibliographical endeavour is to enlarge the understanding that we bring to a reading of geographers' work. It incorporates the sort of detail, frequently biographical in nature, which is of shared interest to all geographers. The way in which geographers have expressed their individuality in diverse ways from such a variety of circumstances provides us with a kaleidoscope of histories and accomplishments. These data invariably derive from archival sources and reveal matters as they really were and not as they might seem to have been. We are reminded continually of the worth of archival materials, deposits extant and those yet to be. Context, too, is significant, though after the facts of the matter have been secured. Yet there are many 'contexts' – the prevailing (macro) context of the continent or country, the context of current geography, and the context of personal, political, or economic situation. It is from amidst this variety of contexts that one begins to understand the work and the worker, the craft and the craftsperson. This undertaking provides us with a new insight with which to view disciplinal accomplishment at different times and in different settings.

It is by chance that five of the authors written about here are American. These were the essays which were submitted. Yet it is especially fitting that these geographers should be written about in a year when the International Geographical Congress is to be held in Washington DC. This is the third time that the USA has been so honoured. The first of these occasions was in 1904, and the second was in 1952. Even prior to 1904 an international geographical congress had been held – in Chicago in

1892 – though this gathering was not of the series commenced in Antwerp in 1871. Three of the five Americans written of would live through the early congresses and two of them would live through the mid-century congress. Four of them had an interest, ranging from keen to very keen, in the history of geographical thought – Pumpelly, Matthes, Glacken and Clark.

Pumpelly lived at a time when geography in North America was beginning to develop its own identity as it was pared away from geology. Pumpelly, unseen mentor to Davis, encouraged the latter in his striving for an evolution of geography. Early in the 1900s Pumpelly took Davis on an expedition to Turkestan, where geology and geography intersected in extended archaeological digs. Round the camp fire Pumpelly had encouraging things to say about the geographical undertaking, although by this time Davis had a geographical discipline already advancing from his physiography in the USA. People of the stature and standing of Pumpelly had much to do with the gradual evolution of geology into a physiography which became recognizably geographical. He had studied at Freiberg in the Royal School of Mines, and brought back knowledge which was to diffuse and have relevance to his work relating to Arizona silver mines, gold, lead and silver mines in Yesso (Hokkaido), Japan, coal-mines in China, and copper mines in Michigan. Between these activities he wrote colourful accounts of his travels which were helpfully geographical in tone and prescription. Pumpelly wrote on the desirability of allowing Chinese to immigrate to the USA, a subject that would later engage Friederich Ratzel's attention.

Unlike Pumpelly, Matthes was born in Amsterdam, the Netherlands, in 1874, and did not arrive in the USA until 1891. Following studies in the Netherlands, Switzerland, Germany, and then the Massachusetts Institute of Technology, he joined the topographical branch of the US Geological Survey. His advancement was swift and sure. In 1900 he published his first scientific work, '*Glacial Sculpture of the Bighorn Mountains, Wyoming*', and in 1907 completed the 'Yosemite Special' map at a scale of 1:24 000, for which he is remembered to the present day. In 1904 he went to Cambridge, Massachusetts, and studied at Harvard under the tutelage of William Morris Davis. It was there that he developed a keen appreciation of things geographical and nearly 30 years later was elected president of the Association of American Geographers. His knowledge of parts of the Western scene with particular reference to topography was without equal. From this interest he traced the emergence of an indigenous American geography upon which he could superimpose the European antecedents and understand the skein of the profession.

Clarence Glacken entered upon a career in geography late in life. At the University of California, Berkeley, he developed his views concerning 'The Idea of the Habitable World', a thesis which he had defended as a doctoral dissertation at Johns Hopkins University (1951). It concerned the relation of man to the physical environment as expressed in intellectual history and the cultural landscape. He recognized persistent themes in the history of thought which have been retained from antiquity to the present. In 1967 his work was brought together in *Traces on the Rhodian Shore*, a volume whose content terminated in the eighteenth century. A sequel to this volume for later decades was never completed though Glacken accumulated a large amount of material which he thought was necessary for the undertaking; this trove remains in Berkeley.

Each of these three geographers in his own way had a deep and abiding sympathy for the need to comprehend the direction in which one's learning was directed. Fulfilment involved on the one hand looking forward towards one's plans and on the other hand looking back through the corridors of time to note the paths which the intellectual pioneers had trod.

John Muir, a Scot removed to the United States when eleven years of age, developed a keen appreciation of place and landscape. He contributed much to what we might call the regional geography of the American West, to glaciology and to an approach to conservation that was in sympathy with things ecological. Inventive and ingenious, he was apparently the first to proffer the notion of a land-bridge between Asia and North America; he also noted the devastation of overgrazing and the virtues of solitude. He described his vision of nature in happy manner with felicity of expression. His writings are evergreen and are being rediscovered in an age when exploitation has all too often meant destruction.

Andrew Clark was born three years prior to the death of Muir. His early life was spent in Canada, attending McMaster University as an undergraduate and the University of Toronto as a postgraduate student, where he was assistant to Griffith Taylor. He then took a doctorate at the University of California, Berkeley (1944), at which time he came under the influence of Carl Sauer. Clark taught in New Zealand, then at Berkeley and Johns Hopkins University (both of these jobs were special posts created for teaching military personnel), and next at Rutgers University before accepting a position at the University of Wisconsin in 1951. Being part of the Wisconsin department in which Richard Hartshorne was a member and having studied under Carl Sauer provided him with a unique position with regard to the posture of historical geography. The culture history of Sauer and the areal differentiation of Hartshorne, viewed as main thrusts in the field, placed Clark at the confluence of powerful moving bodies of thought. Clark adopted the region as his framework for geography: space was the essential. The concept of the historical dimension continually unwinding and continually revealing changes in pattern encouraged him to adopt a mode of sequent occupance. His work was largely empirical and he found it hard to generalize. He studied archival and field sources and dealt with ever smaller areas in ever greater detail. A year spent with H.C. Darby at University College London helped give him cause to ponder the position of historical geography in the pantheon of the geographical schema. He had an interest in the history of historical geography, as witnessed by his chapter on the subject in *American Geography: Inventory and Prospect*; he also had a keen interest in the history of geographical thought.

The three remaining figures in this publication are removed in space or time. Ibn Battuta is revealed as an early traveller, a good writer, an individual with a great variety of interests and curiosities. He was one of the many - and one of the most distinguished - individuals who travelled and explored a largely unknown world, and began to fill the gaps in knowledge. Born in Tangier in 1304, he died in Fez, Morocco, in 1378. He was a religious man, very well educated, who revealed the last glories of the great period of the Islamic empire prior to its invasion and reduction by Mongols, Turks and Spanish Christians. His travels provide us, and provided previous peoples, with a major source of information concerning especially Africa and Asia.

The Polish geographer Stanisław Pawłowski, a former student of E. Romer, contributed much to the field between the two world wars. He had a wide variety of interests, but was especially enthusiastic about the regional concept and geographical education. His studies in the morphology of landscape were also excellent. His many publications, ranging from studies of seashores, glaciers, and the temperature of river waters to anthropogeography and methodology, were read by both schoolchildren and advanced scholars. An archival collection of his correspondence, subject to further excavation, is maintained in the state archives in Poznań.

Finally, Josef Schmithüsen is represented in biobibliographical format. He was a

German geographer whose life spanned most of the twentieth century. He was a craftsman working in the German tradition of regional excellence. Later he became particularly interested in biogeography and in 1972 founded the German periodical *Biogeographica*. He published a handbook concerning *Vegetation Geography* in three editions (1959, 1961 and 1968) and was considered a pioneer in this undertaking. Later he published the *Atlas of Biogeography*, which was his greatest accomplishment. On his sixty-fifth birthday a special edition of his collected works, *Landscape and Vegetation*, was published by his colleagues at the Geographical Institute of Saarbrücken University. This was a splendid recognition of the man's work. He travelled for purposes of research to Africa, Asia, North and South America, Australia and New Zealand, and was intellectually active to the end of his life.

The figures in this publication each made contributions to the development of geographical thought, the strengths and weaknesses of which are not immediately clear. In time this agar will fuse to become part of that larger comprehension which is geography in the making. And so this instalment of *Geographers* constitutes part of an ongoing series which seeks to reveal something of geographers in their quest for the development of geographical science.

Geoffrey J. Martin
Editor

List of Abbreviations

Am. Acad. Arts Sci. Proc. American Academy of Arts and Sciences. Proceedings
Am. Alp. J. American Alpine Journal
Am. For. American Forestry
Am. Geol. American Geologist
Am. J. Sci. American Journal of Science
Am. Rev. Can. Stud. American Review of Canadian Studies
Ann. Assoc. Am. Geogr. Annals of the Association of American Geographers

Bad. Geogr. nad Polską Płn.-Zach.
Bull. Am. Geogr. Soc. Bulletin of the American Geographical Society
Bull. Am. Assoc. Petro. Geol. Bulletin of the American Association of Petrological Geologists
Bull. de la Soc. Belge d'Études Géogr. Bulletin de la Société belge d'études géographiques
Bull. Int. Ac. Sc. et Lettres Bulletin international de l'Académie des sciences et lettres
Bull. Int. Ac. Pol. Sc. et Lettres Bulletin international de l'Académie polonaise des sciences et lettres
Bull. Int. Geod. and Geophys. Union Bulletin of the International Geodesic and Geophysical Union
Bull. Sierra Club Bulletin of the Sierra Club

Can. Geog. Canadian Geography
CIW Carnegie Institution of Washington
C. r. Cong. Int. Géogr. Comptes rendus, Congrès Internationale de Géographie

Deutsche Rundschau für Geogr. Deutsche Rundschau für Geographie
diss. dissertation

Econ. Geogr. Economic Geography
Erdkd. Wiss. Erdkundliches Wissen

Geogr. Rev. Geographical Review
Geogr. Taschenbuch Geographisches Taschenbuch
Geogr. Z. Geografische Zeitschrift

Home Geogr. Mon. Home Geographic Monthly

Int. Geogr. Glossarium

J. Econ. Hist. Journal of Economic History
J. Geol. Journal of Geology
J. Hist. Geogr. Journal of Historical Geography
J. Soil and Water Cons. Journal of Soil and Water Conservation
J. Wash. Acad. Sci. Journal of the Washington Academy of Science

Minn. Geogr. The Minnesota Geographer
Mitt. d. Geogr. Ges. Mitteilungen der Geographischen Gesellschaft
Mitt. d. Geol. Ges. Mitteilungen der Geologischen Gesellschaft

Prace Pozn. Tow. Przyj. Nauk.
Prace Tow. Nauk. Warsz.
Proc. Am. Assoc. Adv. Sci. Proceedings of the American Association for the Advancement of Science
Proc. Boston Soc. Nat. Hist. Proceedings of the Boston Society of Natural History
Proc. Geol. Soc. Am. Proceedings of the Geological Society of America
Prog. Geogr. Progress in Geography
Prog. Human Geogr. Progress in Human Geography

Spraw. Ak. Handl.

Tech. Rev. Technological Review
Terr. Plioc. Pléist.
Trans. Am. Geophys. Union Transactions of the American Geophysical Union

UCLA University of California at Los Angeles
USGS United States Geological Survey

YB Am. Alp. Club Yearbook of the American Alpine Club
YB Assoc. Pac. Coast Geogr. Yearbook of the Association of Pacific Coast Geographers

Ibn Battuta

1304–1378

Akhtar Husain Siddiqi

The Islamic travel literature of the medieval period is a great source of geographical information. Geographical accounts of far-off lands and the observation of strange phenomena kindled interest in people and places. With other Muslim travellers – al-Masudi, al-Muqaddasi, Ibn Hawqal and Idrisi – Ibn Battuta is among the most interesting figures in the history of the Islamic period.

1. Education, Life and Work

Abu Abdallah Muhammad ibn Abdallah ibn Muhammad ibn Ibrahim al-Lawati, known as Ibn Battuta, was born in Tangier, Morocco, in 1304 into a family that had a tradition of judicial service as *qadis* (judges). He belonged to the upper class and received the usual literary and scholastic education of theologians, including a modest share of the sciences (Gibb, 1929, p. 2). Although information concerning his personal life is sparse, his travels provide some idea of his position as a theologian who sought strength through meeting wise men (men of God). Wherever he went he took pleasure in meeting them and learning something he did not know before. His purpose in travel was to meet men of spiritual reality, and this interest is apparent on nearly every page of his work. He died in Fez, Morocco, in 1378 at the age of 74.

Ibn Battuta makes no secret of his private life. He explains how he moved freely in society, how he mixed closely with different classes of Muslims, and how easy it was for him to marry into respectable families. He was a spendthrift, in the habit of spending beyond his means. He testifies to the practice of slavery then in vogue and tells the story of the many slave girls he acquired. A devoted Muslim, Ibn Battuta respected the law-abiding people of other religions as well for their devotion to the principles of their religion. He tells how he was accommodated in rest-houses and hospices, where he was hospitably welcomed and entertained. Fond of music, he liked to travel in the company of singers. Ibn Battuta served as *qadi* at various places. As an envoy of the Sultan of Delhi, he joined the Royal Embassy in China after eight years in the service of Muhammad bin Tughluq as a Grand *qadi* of Delhi.

Husain (1953, p. vii) points out that of all the Arab geographers and historians, Ibn Battuta made remarkable contributions to the history and social conditions of the Muslim communities of the medieval period of the Islamic world, because of his interesting personality and versatile talents as a scholar, theologian, adventurer, warrior, sailor, swimmer, traveller, explorer, pilgrim, botanist, politician, poet, historian, geographer, jurist, ascetic and devotee. Ibn Battuta's travels have been studied by historians and geographers, but no estimate of his work has been satisfactory. Gibb (1929, p. 2) points out that it is first and foremost a human diary, in which the tale of facts is subordinated to the interests and preoccupations of the diarist and his audience. Lee (1829) ascertained the accuracy and fidelity of Ibn Battuta's account and found it worthy of credit. Miquel (1971, p. 725) says that the chronology suffices to demonstrate the new dimensions Ibn Battuta gave to the genre of *rehla* (travel). Instead of offering traditional accounts of the holy places, his travels, in fact, became a description of the known world. This account appears as one of the most typical forms of the literature of observation (*iyan*).

Yule (1966, Vol. IV, p. 12) points out that Ibn Battuta's narrative gives many curious and probably truthful details, such subjects being more congenial to his turn of mind than the correct observation of facts in geography. Fischer, Campbell and Miller (1967, p. 32), recognized him as the 'traveler of Islam'. He saw more of the world than any other Arab traveller of the Middle Ages, as he had an insatiable desire to see any country or place open to him and to meet famous theologians and scholars. They point out that although his point of view was never that of a geographer, and he had little use for maps, he augmented the geographical knowledge of his contemporaries to a considerable extent by being the first to describe for the West many of the countries through which he travelled. Rugoff (1960, p. 21) says that his reputation as a religious scholar smoothed his way as a traveller and in several instances brought him high honours and offices. The more he travelled, the more his motive seems to have been curiosity about peoples and places rather than religious devotion or honours.

Sarton (1948, p. 1616) believes that his account contains a few ambiguities and discontinuities, and that in some cases it is impossible to say, even roughly, how he went from one place to another. Ibn Battuta says, 'It was a habit of mine on my travels never, so far as possible, to retrace any road that I had once traveled over' (Gibb, 1962, p. 283). This habit explains many of the strange groupings of cities in his narrative. Although his point of view was never that of a geographer or of a historian, his account has considerable geographical and historical value.

Beazley (1949, p. 7) asserts that no greater Muslim traveller ever lived than the fourteenth-century doctor of Tangier. The successive invasions and devastations of the Islamic world by Turks, Mongols and Spanish Christians were slowly ruining the higher culture of the Islamic world, and Ibn Battuta's work provides the last glimpses of this empire.

Newton (1926, p. 100) describes Ibn Battuta's narrative as lively and often entertaining; it reveals him as a man of restless energy and curiosity, clear-sightedness and determination to enjoy life. At the same time he was a devout observer of the practices of his religion, with a particular devotion to the saints. Kimble (pp. 102, 125–6) states that Ibn Battuta's narrative is rich in the lore of the times and throws considerable light upon the question of the medieval knowledge of Africa. There is hardly any doubt that Ibn Battuta's travels serve as a major source of information about the contemporary political, social and economic conditions of a large region of Asia and Africa.

This review provides hardly any appreciation of his work as a geographer. The purpose of this study is to investigate the travels of Ibn Battuta and the geographical phenomena he encountered, and to understand the extent to which his perception of these phenomena was conditioned by the particular spatial-temporal context in which the travel accounts were produced. These accounts provide information about the geographical and historical context in which they were presented, despite the fact that the framework of *rehla* (travel) does not provide a realistic perspective on the nature of geographical enquiry in the Middle Ages. The goal of this study is to stimulate further interest in and encourage critical examination of these travel accounts for a diverse audience of geographers.

Ibn Khaldun (pp. 307–8) advanced the opinion that a scholar's education is greatly improved by travelling in quest of knowledge and meeting the authoritative teachers (*shaykhs*) of his time. The reason, he argued, is that human beings obtain their knowledge, character, opinions and virtues either through study, instruction and lectures, or through imitation of a teacher and personal contact with him. The only difference is that habits acquired through personal contact with a teacher are more strongly and firmly rooted. Thus, travelling in quest of knowledge, meeting authoritative teachers and establishing contact with scholars are absolutely necessary for the acquisition of useful knowledge and perfection. Ibn Khaldun gave an eye-witness account of Ibn Battuta's reception at the court of Sultan Abu Inan and discussed the scepticism generated by his stories. However, he did not say anything about him as a man.

Ibn Juzayy put Ibn Battuta's account into literary style by command of Sultan Abu Inan. In the introduction, he reports that Ibn Battuta made the pilgrimage to Mecca as the fulfilment of one of the duties imposed by his faith, but also as an opportunity to broaden his education and enhance his qualifications for judicial office by contact and study with famous scholars (Gibb, 1958, p. x). Ibn Battuta says that he attended lectures at Damascus, that scholars granted him a licence to teach (Gibb, 1958, pp. 154–6), and that diplomas were conferred on him; it would seem that this was, indeed, the limit of Ibn Battuta's original purpose. Instead of remaining in Mecca to study, he spent the following years in further exploration of the lands of classical Islamic culture.

Ibn Battuta did not cite any historical and literary sources. He narrated things as he saw them. He did mention natural and anthropological sites, as well as the towns or cities of great commercial importance. It is hard to say whether he profited by contacting merchants who came from distant countries. His knowledge of geography, as compared with that of other Arab geographers, seems to have been imperfect. Ibn Battuta is not overly concerned with chronology. His dates for some of his journeys cannot be reconciled with some of the facts that he mentions, but his grasp of contemporary history is generally borne out by historians. His account contains valuable information about Byzantine, Muslim rulers, and the Asian and African countries he visited. He relied on firsthand knowledge by means of contact with those in the service of states. Hence he is a careful observer who combines social and political insights. The material presented is by no means confined to the Arab world but includes non-Muslim regions as far as he could obtain information.

Yule (1966, Vol. II, pp. 24–5) remarks that it is singular that the narratives of Marco Polo, Odoric, Niccolò Conti and Ibn Battuta, the four most remarkable

Asiatic itinerants of the Middle Ages, should all have come down to us under the disadvantage of dictation. Ibn Juzayy (Gibb, 1958, pp. 7–8) says that he rendered the sense of the narrative of Abu Abdallah (Ibn Battuta) in language that adequately expressed the purpose he had in mind and set forth clearly the ends he had in view. Frequently Ibn Juzayy reported Ibn Battuta's words in his own phrasing. Ibn Juzayy further says that he included all the anecdotes and historical narratives that Ibn Battuta related, without applying himself to investigate their truthfulness or to test them, since Ibn Battuta himself adopted the soundest methods of authentication and disclaimed responsibility for the more doubtful by warning expressions. Names and places that offer difficulties were assisted by means of vowel signs and diacritical marks to have a proper pronunciation; non-Arabic names were explained to avoid confusion.

Like other Muslim scholars, Ibn Battuta had a remarkable power of memory. In his case the literary and religious importance of the oral transmission of knowledge could be closely observed. The memorizing of the Qur'an, and vast quantities of traditions as well as poems and stories, in which Ibn Battuta excelled, is indeed a noteworthy fact. Ibn Battuta admitted that during his adventures he sometimes lost the notes made on his travels, and then rewrote them from memory; it is probably not unusual. Some inaccuracies may be due to the ambiguities of Arabic writing. In the oral transmission of a text, errors of this kind are possible. Other difficulties were aggravated by the fact that his work transcended the boundaries of languages and civilizations. Arabic translations of Sanskrit and Persian words produced textual corruption, due to the problems of expressing foreign sounds. Words in a foreign language that are quoted in an Arabic context sometimes present great difficulties to the reader.

IBN BATTUTA'S TRAVELS

Ibn Battuta started his wanderings at the age of twenty-one (1325), and did not cease until he was fifty-one (1355). He commenced his travels by traversing the whole length of North Africa from Tangier to Alexandria, pausing at Cairo (1326) and the valley of the Nile, and proceeding to visit Palestine and the rest of Syria, including Aleppo and Damascus (1326). He then performed the pilgrimage and went to Meshed. From there, he went to Basra, through Khuzistan to Isfahan, to Shiraz and back to Baghdad. After visiting Mosul, he made the pilgrimage for a second time. He made a voyage down the Red Sea to Yemen, through which he travelled to Aden. From Aden, Ibn Battuta continued down the East African coast. He travelled through Africa from the north to the south and then from the east to the north-east; and the information that he supplied has been found consistent at all points with the accounts of modern explorers. From here he sailed to the coast of Oman and proceeded to Hormuz. After visiting Kish, he crossed the Gulf of Bahrain, al-Hasa and thence crossed central Arabia, and made the pilgrimage again.

Later, he revisited Syria, and made an extensive journey through the Turkish Sultanate. He then crossed the Black Sea and visited Constantinople for the first time. He also visited a thermal spring at the foot of the Caucasus and found a guide to conduct him to the city of Bolghar. After a short stay at Constantinople, he went back to Uzbek and thence took his way across the desert to Khwaris and Bukhara. His travels then extended through Khorasan and Kabul, including a passage of the Hindu Kush. He then proceeded to Sind, reaching the Indus in 1333. After visiting the Indus Delta he went northward to Multān and on to Delhi (1334). He travelled through India, the Maldives, Ceylon, Sumatra and China. On his return from

China he came to Calicut (India) and then visited Dhofar, Muscat, Shiraz, Isfahan, Basra, Najaf and Baghdad. He made his seventh and last pilgrimage in 1348, returned to Cairo and then started for Morocco, reaching Fez in 1349. After some time he set out in the direction of Spain, visited Gibraltar and Granada, and returned to Fez. After some days he set out again and proceeded in the direction of the Niger. In 1353 Ibn Battuta left Mali, exploring the middle course of the Niger (which he called the Nile of the Negroes). He visited Timbuktu and other places and finally arrived at Fez.

Negotiating rapid streams, burning deserts, treacherous ravines, dusty roads, fertile lands and rough, dreadful seas, Ibn Battuta covered about 78 000 miles in the course of his long travels. He visited all the places of importance in the Islamic world and the court of every Muslim ruler of his age. He also visited the non-Muslim countries of Constantinople, Ceylon and China.

It was natural that his travel account should hold the interest of his audience. As we read Ibn Battuta's account, we become increasingly aware of the fact that he is trying to provide something other than mere entertainment. Obviously, he was presenting his tales of experience to his readers, young or old, in order to influence their character through moral precepts by providing anecdotes. He began as a young man, knowing nothing about the world or the people, with the sole purpose of pilgrimage. He never engaged in trading ventures. His accounts give evidence that he was not interested in matters of profit or loss, which are the interests of merchants. With time and experience, he gained maturity in dealing with people.

Ibn Battuta's interest in the cultural landscape of the Islamic world was catholic in nature, and his awareness of Islamic cultural centres was wide and varied. He began his career within the restrictive confines of a particular school of religious thought. His travels emphasized his desire to meet religious authorities, but he was observant enough to see the actual Islamic world in which people lived by (a) exploring territories and landscapes and (b) establishing personal contact with the people and their way of life. Thus, in his travels he faced varied environments, customs and foods, meeting diverse people bound by a single religion (Islam).

2. Scientific Ideas and Geographical Thought

Ibn Battuta possessed an awareness of and curiosity about the character of places and Muslim communities. He catalogued information on locations, places and products. He emphasized Islamic culture rather than the physical environment in which people lived. He provided information on the behaviour and the common life of a group of people. He looked into similarities or contrasts in speech, behaviour, ideology, livelihood, value systems and societies in which people lived together in a totally Islamic culture. He also observed spatial variations among cultural groups and the functioning of society, and he described and reflected on the ways language, religion, economy, government and other cultural phenomena varied from one place to another.

CULTURAL HISTORY

His cultural region was a uniform area inhabited by people who had a single religion (cultural trait) in common. Although the groups shared the same religious beliefs, one can identify an infinite number of formal cultural regions from the description of different areas he visited. In the study of culture, one of the most important questions to be asked is what level of detail is appropriate for investigation. Ibn Battuta observed a large range of cultural groups, and his units of observation were extremely restricted, so it is hard to draw broad conclusions. Although these groups were similar in many ways, enough differences existed between the Arabs and non-Arab groups for them to be considered separate cultural units. Organized communication provided a means of maintaining orderly relations with other people over long distances. A network of travellers existed, passing from one Islamic city to another by land, river or sea.

Ibn Battuta's travels reveal that the Islamic world was divided into subregions, centring on the Persian Gulf, Egypt, North Africa, East Africa, Syria, Iraq, Asia Minor, the central Asian steppes and western India. One can see the development of commercial intercourse among these regions and adjacent areas. The commerce that flowed between the subregions provided the links that bound them into a world system. On the one hand, the Islamic world had created co-operative linkages with other areas; on the other hand, each area was independent. Each area benefited from co-existence and mutual tolerance.

Ibn Battuta tried to explore the cultural and spiritual aspects of the Islamic world. In his opinion a community of Islam had a social obligation to know about other societies or elements of society, whether they are good or evil; the production of new knowledge was a way of understanding the will of God. Therefore, travel must be understood as a general activity and not as an elitist enterprise. He outlined the power and wealth of the local rulers and also described the conditions of the ordinary people. For example, Ibn Battuta tells us, the 'people of Khawarizm are well-to-do, have reception halls, furnished with rich carpets, their walls hung with cloth, and use vessels of silver-gilt and Iraqi glass with wooden spoons' (Gibb, 1971, p. 545). He communicated with people of different status and sought information. These are by no means small achievements of his travels. He tried to describe various social and cultural segments of the societies he visited. Since there is a need to understand the social character of the societies and appreciate social and political mechanisms, his accounts ensure that Ibn Battuta gets the recognition he deserves. His travels provide information about the spiritual, cultural and environmental needs of Muslim societies. His descriptions establish the true nature of the places he visited, but it is easy to see that his observations were selected and biased, as well as coloured by the perception of a storyteller. At an individual level, his selection of narration is based on value judgements, political considerations and personal preference. Thus, his travel accounts allow one to distinguish the social circumstances in which he visited these places. Based on his account, one can trace the social and cultural milieu of Islamic society and reconstruct the history and the social and material culture of his time.

Thus, to Ibn Battuta, places to visit were unique, offering him the culture of the area. The study of people associated with places, especially the centres of religious importance, as well as the experience of different people, provided him with a sense of the direction of Islamic culture. To present-day geographers, such literature has also been of value as they search for an understanding of the unique character of places in history and their evolution.

POLITICAL ORGANIZATION

Ibn Battuta did not find political unity in the Islamic world, but he did find an Arab cultural unity and the sense of belonging to one larger community, ruled by Quranic law. The Arab ruling groups were absorbed by the lands they conquered such as Persia and India. The Islamic world was recognizable in art, architecture, mosques, caravanserais, hospitals, schools and aqueducts. The great Arab cities played a special role as centres of learning as well as trade. Schools drew students from all over the Islamic world, developing a rich tradition of religious and legal learning. Ibn Battuta described an Islamic system of social morality and law related to the principles of social justice inherent in the Qur'an. The order of *ulema* (scholars) was held sacred throughout the Islamic world, and the *faqhi* (interpreters of religious law), the *qadis*, the *khatib* (readers of the Qur'an), the *sufi*, and the *shaykh* (religious scholars) were uniformly revered.

Ibn Battuta did not travel to modify his concept of the physical world in which he lived. He visited a large number of towns, cities and villages and consistently used the same phraseology to identify or describe them. He did discuss physical conditions in some detail, and he associated the condition of travel with particular historical events that had created the social and cultural or material conditions of the area. On his way through central Asia, he found places (such as Bukhara) that had been plundered and destroyed and had not recovered. Surprisingly, Ibn Battuta did not mention any of the detailed scholarly geographical treatises that were available in this period.

SOCIAL VALUE SYSTEM

He knew that the Islamic world stretched from the Atlantic Ocean to China, and he visited it all. He knew that as a Muslim he could confidently expect generous hospitality from his co-religionists, especially if he had any reputation for piety or religious scholarship. Ibn Battuta tells us how on his arrival *qadis*, missionaries, and governors and Muslim merchants came out to receive him, a custom of Muslim society during the Middle Ages. To serve the Islamic educational and scholarly institutions was the law of *waqf* (pious endowment), which allowed for the establishment of sinecures for teachers and students to pursue their studies. The student–teacher relationship was one of the central ties of Islamic society; educational institutions (*madrasas*) were places to study *fiqh* (jurisprudence) and to advance other kinds of knowledge or intellectual endeavour in order to co-ordinate their usefulness to *fiqh* inquiries (Makdisi, 1981). Ibn Battuta did not mince words when he saw the slave market in Turkey. He found it culturally, socially and spiritually disastrous, as he saw that people were corrupted by the officials of the place.

COMMUNICATION SYSTEM

The complex system of trade had established routes that passed through the heartland of Islam and provided passage to the East and West. The northern route from Constantinople connected the places in central Asia. The central route connected the Mediterranean with the Indian Ocean via Baghdad, Basra and the Persian Gulf, while the southerly route linked Alexandria, Cairo and the Red Sea with the Arabian Sea and the Indian Ocean. In addition, secondary routes extended in all directions. Ibn Battuta points out the rapid communication between the capitals and important towns and the outlying provinces, and the elaborate post system that was kept up. Every few miles there was a postal station where official passengers could get a fresh team of horses or pass on their dispatches to another member of the same service. He also talks about dispatching carrier pigeons with news. A network of roads was kept in good order to enable rapid transit between administrative centres and strategic points. The road system stimulated travel as well as commerce. Ibn Battuta tells us that because of disturbed conditions in a number of areas he followed safe routes under the protection of archers and horsemen, describing the disorderly conduct of the Bedouin who robbed people on highways.

URBAN ENVIRONMENT

In his travel accounts, Ibn Battuta describes the towns located along the canals, rivers and coasts. He speaks of markets and wealthy merchants dealing in manufactured goods, which were exported to different parts of the Islamic world. His descriptions of cities and towns are generally rhetorical with a few facts about goods and commodities found in the bazaars. He repeatedly described delightful gardens with trees and the choicest fruits, a beautiful landscape and the rivers flowing with crystal waters. In fact, he was repeating Quranic verses that show heavenly life, which have been repeated in many ways (Qur'an, S. II, 25).

Ibn Battuta's account shows that each city, in addition to its role in the Islamic world system, served as the centre of its own region, implementing the local trade in common foodstuffs and goods of everyday use. Each city-region was made up of a multiplicity of local units; only a very small proportion of their activities had any influence beyond their particular urban environment.

Ibn Battuta often talks about the gardens and orchards along his route. Gardening on a large scale developed during the rise of Islamic civilization. These gardens were both functional, providing shade and agricultural products, and aesthetically pleasing environments in the dry climate of the Islamic world. On the other hand, orchards, vineyards and groves reflected the constraints of land and water, as people began to export products for long distances from their gardens. These fruits were used in great abundance in social and religious festivities. Ibn Battuta looks at gardens as a manifestation of the cultural views of the Islamic era, as well as a divine gift of the natural environment. Thus, one can speculate that the rural and urban orientation of the respective culture systems was reflected in people's relationship with nature as expressed in their gardens.

COMMERCIAL PRODUCTS AND TRADE

Ibn Battuta provides information about Indian trees, fruits, grains and spices. He talks about olive cultivation in Syria and the rich fruits (figs, raisins and olives) produced at Sidon. He describes the cultivation of bananas, betel-nut and *pawn* (betel leaf). He talks about oranges, grapes and pomegranates, and he describes cocoa palm, areca palm, cloves, Indian aloe, jack tree and camphor cane on Java. He provides an account of the coconut and its uses. He reports that the melons of Khawarizm have no equal except the melons of Bukhara and Isfahan. He describes agricultural produce in a colourful language, and associates its use with local tradition and custom.

He provides an account of obtaining pearls and the processes that produce them, and describes the manufacture of cloth, the linen industry and the manufacture of excellent woollen fabrics in Egypt. He describes the sugar-cane presses located close to the Nile. He tells us that the upland region of Syria was predominantly an olive-growing region, manufacturing soap and fine cotton fabric. He found Nablus rich in olives and olive oil. He talks about Ladhiq's (Anatolia) manufacture of cotton fabric edged with gold embroidery, and he describes the wooden vessels and spoon-making industry of Al-Blabakk (Syria).

He found the vicinity of the Tigris near Al-Qayyara abounding in pitch; he noted seepages of bitumen and found similar substances between Kufa and Basra. Ibn Battuta tells us, 'Pitch (like black clay) is found between Kufa and Basra: it was shovelled up from there and transported to Baghdad' (Gibb, 1961, p. 347). At Al-Qayyara near the Tigris, 'there is a patch of black land in which are springs that exude pitch, and tanks are constructed for it to collect in' (Gibb, 1961, pp. 329–30).

Besides the products of daily use, trade included goods of high value, which were in demand by the wealthy, privileged and powerful class. One gets the impression that the population in general was living a self-sufficient life, along with an incipient bourgeois commercial class.

Trade was the driving force of the Islamic world, and trade was, by and large, concentrated in cities and conducted by city merchants. Each city constituted, as it were, an oasis, surrounded by its own back country, its periphery.

MATERIAL CULTURE

Architecture is not only a stage in modernization in a given era; it also reflects the level of technological skills available to the architect. Ibn Battuta provides some descriptions of architecture: 'The city of Alexandria . . . is remarkable in appearance and solid of construction . . . and of memorable edifices both secular and religious. Noble are her dwellings, graceful her qualities, and to imposing size her buildings unite architectural perfection' (Gibb, 1958, p. 18). In Syria he describes Hebron as 'a town of small extent but high distinction and shining splendours, pleasant in outward appearance, lying in the hollow of a valley' (Gibb, 1958, p. 73). He tells us that Shiraz is a solidly built city with elegant gardens, gushing streams and handsome thoroughfares (Gibb, 1962, p. 299). He found the city of Al-Mawsil, on the bank of the Tigris, and its citadel of solid and imposing construction, strengthened by massive towers. In the court of the mosque he found a cupola; inside was an octagonal basin of marble supported by a marble column. Water spouted out of this basin with impetus, rose to the height of a man, and then fell, providing a beautiful spectacle. In the dry parts of West Africa and Egypt, where there are no springs and rivers, towns used to collect their water supply in cisterns, draining the water from large plastered surfaces. Ibn Battuta identified many locations (such as those north of the Caspian Sea) where he found known water sources, surface pools (of rain) and shallow wells, providing the halting places in the desert country.

Ibn Battuta visited castles, baths, fortified palaces and monuments. He saw art and architecture rooted in the Greco-Romano-Byzantine traditions of the eastern Mediterranean world. These buildings were once thought to be the desert retreats of Umayyad princes and noblemen. In many areas he found skilled hydraulic engineering, through which people harnessed scarce desert water resources. The fortresses, the palaces, the mosques and the extensive agricultural facilities located in the middle of the desert were exciting to visit. The areas he visited reflected a rich and dynamic culture. The hinterlands of Syria, Iraq, central Asia and Persia were developed to exploit the agriculture and trade potential of what, in the pre-Islamic era, had been the marginal frontier region. These areas provided a broader understanding of the total environmental and economic context.

Many places were self-contained economic units, based on water conservation and storage, agriculture and trade. Future historical geographers may study these integrated agricultural complexes and determine the full political and cultural context in which they were developed.

3. Influence and Spread of Ideas

To a careful reader of Ibn Battuta's travels, there is no shortage of ideas and information on a host of subjects that a historical and cultural geographer would like to know about his age. Ibn Battuta was one of the travellers who had a desire to visit religious places and discuss *fiqh* with scholars. A large portion of his narrative is related to these things. Yet a claim can be made for Ibn Battuta as one of the cultural historians of his time. Although there is much about his travel account that can put off the modern reader (Haig, 1887; Janicsek, 1929), Ibn Battuta's religious narratives provide information that should fascinate any one interested in the Islamic period of the Middle Ages and seeking information on the period's social, economic, moral, political and administrative conditions.

One way of trying to understand Ibn Battuta's age is to ask what sort of people he travelled with or what places he visited. As a young man, Ibn Battuta tried to look at the world more closely than anyone before him, and he then used his experiences and observations to describe the different parts of the world in terms of cities, commerce, customs and traditions, architecture, and wealth and prosperity.

Ibn Battuta, undoubtedly, had a hunger to visit new places. On the other hand, he saw the tyranny of the sultans and their agents, famine in India, and plague in Syria, all of which made him unhappy. He always liked products that were well-made and artfully designed, whose value in his mind was measured by the ability of the user to enjoy them, and which therefore enhanced Islamic civilization in industrial production and trade.

In short, Ibn Battuta tried to recollect his experiences, providing the sociocultural features of all the regions he visited. He makes a good attempt to package them as one early sociocultural historian's account, and he manages to produce them without diminishing his greatness as a traveller.

Bibliography and Sources

Ali, A. Yusuf (trans.), *The Holy Qur'an*, Indianapolis, IN: American Trust Publication, 1977.

Beazley, C.R., *The Dawn of Modern Geography*, Vol. II, New York: Peter Smith, 1949.

Fischer, E., R.D. Campbell and E.S. Miller, *A Question of Place: The Development of Geographic Thought*, Arlington, VA: Beatty, 1967.

Gibb, H.A.R. (trans.), *Ibn Battuta: Travels in Asia and Africa*, London: Broadway House, 1929.

—— (trans.), *The Travels of Ibn Battuta*, Vol. I, Cambridge: The University Press, 1958.

—— (trans.), *The Travels of Ibn Battuta*, Vol. II, Cambridge: The University Press, 1962.

—— *The Travels of Ibn Battuta*, Vol. III, Cambridge: The University Press, 1971.

Haig, M.R., 'Ibn Battuta in Sindh', *Journal of the Royal Asiatic Society*, New Series, Vol. XIX, 1887, pp. 393–412.

Husain, Mahdi, *The Rehla of Ibn Battuta*, Baroda: Oriental Institute, 1953.

Janicsck, S., 'Ibn Battuta's journey to Bulghar: Is it a fabrication?', *Journal of Royal Asiatic Society*, 1929, pp. 791–800.

Khaldun, Ibn, *The Muqaddimah*, translated by F. Rosenthal, Vol. 3, New York: Pantheon Books, 1958.

Kimble, G.H.T., *Geography in the Middle Ages*, New York: Russell and Russell, 1968.

Lee, Samuel, *The Travels of Ibn Battuta*, London: J. Murray, 1829.

Makdisi, George, *The Rise of College: Institutions of Learning in Islam and the West*, Edinburgh: Edinburgh University Press, 1981.

Miquel, A., 'Ibn Battuta', *Encyclopaedia of Islam*, New Edition, Vol. III, H-IRAH, Leiden: E.J. Brill, 1971, pp. 735–6.

Newton, A.P., *Travel and Travellers of the Middle Ages*, New York: Alfred A. Knopf, 1926.

Rugoff, Milton, *The Great Travelers*, Vol. 1, New York: Simon and Schuster, 1960.

Sarton, George, *Introduction to the History of Science*, Vol. III, Baltimore, MD: Williams and Wilkins, 1948.

Yule, Henry, *Cathay and the Way Thither*, Vols I–IV, *Ibn Battuta's Travels in Bengal and China*, 2nd Series, New York: Paragon Book Gallery, 1966.

Akhtar Husain Siddiqi is Professor of Geography, Indiana State University, Terre Haute, Indiana, USA.

Chronology

25	February	1304	Born in Tangier (Morocco)
12	June	1325	Left Tangier for Mecca (pilgrimage)
5	April	1326	Visited Alexandria
31	July	1326	Left Cairo for Syria
9	August	1326	Visited Damascus
7	November	1326	Performed first pilgrimage (Mecca)
7	May	1327	Visited Isfahan
	June	1327	Visited Baghdad
27	October	1327	Performed 2nd pilgrimage
26	September	1330	Left Mecca and journeyed through East Africa and some of the Persian Gulf and back to Mecca area
	October	1330	Left Mecca via Jedda, passed
	June	1331	through Egypt, went to Asia Minor and landed at Alaya
26	June	1332	Visited Ladhiq (Anatolia)
18	September	1332	Visited Constantinople
	Early January	1333	A short stay at al-Sara (central Asia)
	Mid March	1333	Visited Bukhara
12	September	1333	Crossed the Indus River
5	November	1333	Visited Multān

20	March	1334	Reached Delhi
8	June	1334	Met Sultan Mohammad
	June	1341	Left the service of Sultan
	August	1341	Appointed ambassador to the King of China
2	January	1343	1st visit to Calicut
9	June	1344	Visited Maldive Islands
26	August	1344	1st visit to Ceylon
13	October	1345	Visited Colomb
9	July	1346	Embarked on board a ship bound for Sumatra
	?	1346	Visited China
	January	1347	Left China via Sumatra and Calicut
	April	1347	Arrived at Dhofar (Zafar) (Arabia)
	January	1348	Reached Baghdad
15	November	1348	Came to Mecca
1	March	1349	Performed last pilgrimage
13	March	1349	Left Mecca for Egypt
	June	1349	Arrived at Tunis
	August	1349	Left Tunis for Sardinia
8	November	1349	Arrived at Fez
		1350–51	Left for Spain and journeyed through Gibraltar, Granada and back to Fez
18	February	1352	Set out to Negroland
28	June	1352	Arrived at Mali
29	December	1353	Left Sijilmasa for Fez
9	December	1355	Finished dictating his travel accounts
		1377–78	Died in some Moroccan town (probably Fez)

Andrew Hill Clark

1911–1975

David Ward and
Michael Solot

Andrew Clark embodies, paradoxically, some of the main cross-currents of human geography during the decades before and after his death. On the one hand, he held firm to a conviction that the most revealing insights of geography were its contributions to the humanities. The simple yet profound human curiosity about what places are like was the only justification he thought necessary for the discipline, and he considered it a privilege to be able to devote his professional life to its pursuit. On the other hand, Clark pursued this commitment with a kind of hyperempiricism that left little room for the evocation of landscapes and broke regional wholes into pieces that could be described in exhaustive detail. In his major works, and, in particular, in the monographs on Maritime Canada, an abundance of maps and tables portrayed almost every aspect of the documentary record, while his text rarely strayed into speculation, comparison or generalization. The result was a carefully crafted exhumation of primary sources from which the imagination and vitality that Clark displayed to those who knew him emerged only occasionally. In contrast, his first publication, a sketch of Prince Edward Island, and his final essays on Canadian landscapes exemplify his humane concern with and his sensitivity to regional identities. In his teaching and advising he sought to sustain this vision in a period when hypothesis testing, quantification, and concentration upon ever more narrow topics susceptible to these methods were becoming standard in the field.

For Clark, geography was a vocation that began with a deep interest in regions as wholes, whose essence, he believed, could be captured only by patiently recovering their changing geographies through time. As the leading historical geographer of his generation during the 1960s and early 1970s, he vigorously promoted his views, but he found himself increasingly marginalized in a discipline that had been lured away from both culture and history by the dubious scientism of spatial analysis. Nevertheless, as the supervisor of no fewer than *nineteen* doctoral candidates, some of whom were themselves to make major contributions to historical geography, Clark was clearly recognized as one of the finest postgraduate mentors in North American geography. His seminar in historical geography at the University of Wisconsin-Madison was open-ended in its topical range but demanding in its intellectual discipline. Postgraduate students in history were frequent participants in his seminar, and Clark unquestionably established the reputation of historical geography among American social and economic historians. In his capacity as mentor, he encouraged the imaginative elaboration of historiographical issues in specific regional contexts, but only in the last decade of his life did he himself reveal, in a series of humanistic addresses, his own imaginative vision of historical geography. This vision was in part an existential geography that reached back into his family past.

1. Education, Life and Work

Clark's personal roots in the Maritimes ran deep. He liked to tell the story of one of his ancestors from the southern Highlands of Scotland who was press-ganged into the British navy during the Seven Years War, and who escaped in Quebec, eventually to settle on Prince Edward Island and establish a connection that persisted into the twentieth century. Clark's parents were born there and his grandparents still lived there when he was a child, and it was over the course of several summers spent visiting them as a boy that he formed his first and most enduring attachment to the island. He was not born there himself, however. His father had left to work as a government surgeon on an Indian Reserve in Fairford, Manitoba, where Clark was born on 29 April 1911. Four years later the family moved south to Brandon, where the elder Clark practised in the town's large Polish and Ukrainian communities, as well as serving as chairman of the board of deacons of the Baptist church and frequently as president of the local Prohibition society. The experience of growing up in relatively peripheral areas in close contact with recent immigrants from undeveloped parts of the Old World fostered in Clark a deep curiosity about the cultural consequences of modernization. Unlike Carl Sauer, his Berkeley mentor, he was neither a 'romanticist' nor a 'transcendentalist', but he espoused a kind of 'Whiggish' belief in the enduring benefits of tradition in the inevitable course of modernization.

If family history provided Clark with a firm basis for his career as a historical geographer, his early training did not. After graduation from secondary school in Brandon and course work at the local college, Clark studied mathematics and physics at McMaster University, and was awarded the BA in 1930. His only overtly geographical experience at this time was his employment during summers as a surveyor's assistant for the provincial highway authority. A year of postgraduate training in statistics at the University of Manitoba in Winnipeg led to a job as a statistician for a Toronto-based life insurance company for four years beginning in 1932.

Clark's postgraduate education began in earnest at the University of Toronto in 1935 when he took up the study of economic history, at first based on his statistical background, under the direction of Harold Innis. Clark continued working as a surveyor during summers, now for the Canadian Geological Survey in upper Manitoba and Quebec. His first academic appointment was an instructor's position in the university's new department of geography, established with Griffith Taylor as its sole professor. Clark worked with Innis and Taylor until 1938 when he received an MA by examination in geology, economic history, and geography. His study of physical geography with Taylor, including accompanying him on a trans-Saharan expedition, cemented his profound sense of environmental influences. Although Clark never accepted Taylor's geographical determinism, his historical work was always based on a thorough understanding of the physical context, much of it gained at first hand. He regarded an environmental sensitivity as indispensable to human geography and he took a real delight in it as well, teaching introductory courses in physical geography until late in his career.

More congenial and formative, however, were the personal influence and work of Innis. Perhaps best described as a political economist, Innis held the title of professor of economic geography ('signally appropriate', as Clark noted in his obituary) at Toronto in the early 1930s when he completed his pioneering studies of the role of primary industries in Canadian economic development. His ideas on the regional concentration of staple export production draining vast extractive hinterlands were, along with the frontier hypothesis of Frederick Jackson Turner, the only theories Clark considered relevant to his historical geography of Maritime Canada, though in the end he found them unsatisfactory. Clark took from Innis a sensitive understanding of the role of the physical environment in economic history, a recognition of the significance of peripheral areas in Canadian development, and above all a demonstration of the value of historical documents in regional research and a model of what their mastery could reveal.

Innis arranged for Clark to transfer to Berkeley in 1938, with a teaching fellowship, to pursue the Ph.D. under Carl Sauer. He began fieldwork for a thesis on Prince Edward Island in the summer of 1939, but after being advanced to doctoral candidacy in 1940 he was temporarily deflected from research on the Maritimes by the offer of a teaching post in geography at Canterbury University in New Zealand. During his two-year stay in the antipodes he completed research for a dissertation on the introduction, by Europeans, of plants and animals foreign to the South Island and their effect on the development of its rural economy. This study, with its attention to the ecological role of specific material culture traits in the creation of a rural landscape, bore the strong imprint of Sauer and his school. The impetus for the thesis (which was published as a book in 1949) was Clark's realization that the re-creation of a British landscape by immigrants to New Zealand, though accepted as a fact by most observers, was itself something that could be put to the test, and was a question well suited to his temperament as a geographer.

From this perspective the thesis is a classic Berkeley work, concerned with the diffusion of particular traits from a hearth area and their role in making over a rural landscape. Clark aimed to show how neither environmental nor cultural factors had foreordained the re-creation of a British landscape in the antipodes. He found instead a lively interplay of culture and environment with a variety of accidents of timing, climate, propinquity, and technology that gave an increasingly British cast to the landscape of South Island as the nineteenth century came to a close. From another perspective, however, Clark's emphasis on the *region* as an organizing theme, and his concern with describing the rise of a nation completely *sui generis*, set his work apart from Sauer, who, in his mature work at least, did not define his problems in political or regional terms. Clark conceived his thesis from the beginning as a regional monograph, and eventually came to see it as part of a long-term research agenda addressing the blending of nature and culture in the major areas of British overseas settlement in the mid-latitudes.

Such academic work, however, was temporarily interrupted by the war. Clark and his wife Louise, whom he had met at Berkeley, first volunteered for service in the American Red Cross in Australia, but they returned to California at the end of 1942. Clark then took a position in Berkeley as a lecturer in meteorology for the Air Force, which was followed by another lectureship for the Army in Baltimore. There he taught the geography of Italy as part of a group of geographers doing military intelligence work under Isaiah Bowman at Johns Hopkins. In 1944 Clark joined the Office of Strategic Services (OSS) in the Research and Analysis Branch of the Far East division as an expert on Australia, New Zealand and New Guinea. He took pleasure in pointing out that in two and one-half years in the OSS he never spent a minute on any of those areas, which was why, he speculated, he was offered the New Zealand desk in the State Department at the end of the war.

Clark considered most of his work for the OSS a futile exercise in a large and poorly organized bureaucracy, but when he was sent into China just as the war ended he used the opportunity to learn as much as possible about that country. Eventually he found himself in charge of the liquidation of the OSS office in Chungking. He took the licence his position granted to arrange three interviews with Chou En-lai, who was then representing the Communists in the city. Although Chou impressed him as articulate and humane, as a practical matter Clark never considered that the USA had any choice but to support Chiang Kai-shek and the Kuomintang. He was under no illusions about their corruption, inefficiency, and lack of popular support, but he instinctively saw Stalinist communism as a global threat to freedom, and worried after the war about 'all of Asia . . . falling under second-hand domination from Moscow'.

Clark was working at Rutgers during this period. After turning down a post in the State Department just after the war, he came to the university in 1946 to begin the first geography instruction in its history. In 1949 he was named chairman of Rutgers' first department of geography, President Robert Clothier citing the significance of the field 'in a world which an airplane can circle in a few days, and in which our country has become the chief trustee of human freedom'. By the time Clark resigned two years later, his energy, enthusiasm and administrative skill had built a small but vigorous department with instruction in all of the main subfields and an ongoing programme of research.

His own research programme included several articles on New Zealand, leading to the publication of his dissertation as *The Invasion of New Zealand by People, Plants and Animals: The South Island* (1949). He also began to redirect his attention to Canada and the Maritimes with a major review essay and other exploratory pieces, which, by mid-century, had solidly established his reputation in the historical geography of Anglo-America. The leading figure in the field at this time, Ralph H. Brown, was primarily a geographer of the mountain West and the Great Plains, though he had produced an eclectic series of studies on North America and a cross-sectional snapshot of the Atlantic seaboard round the turn of the eighteenth century. Brown had worked alone, however, without antecedents, but Clark clearly attempted to recognize him in this emerging vision of historical geography.

As a product of the Berkeley department and as a geographer with experience in the Army and the OSS, Clark was more suited to a position in the mainstream of American geography, and after the early death of Brown in 1948 he emerged as the chief spokesman for historical geography in the USA. Two universities, Wisconsin and Toronto, made offers to him in 1951, and he settled on the University of Wisconsin at Madison, in part because it involved fewer administrative duties. As a matter of fact, Clark was a dedicated academic citizen and served as chairman of the department from 1957 to 1961. He was highly regarded for his service on university committees and his willingness to make tough judgements on academic policies.

This appointment at Wisconsin, at a time when it was regarded, along with Berkeley and Chicago, as one of the leading programmes in North America, gave him the opportunity to provide a bridge, of sorts, in what he called the 'mild debate' between Carl Sauer and Richard Hartshorne concerning historical geography. As a student of the former and now colleague of the latter, he sought to reconcile Hartshorne's limited view of the role of history in geographical practice in *The Nature of Geography* (1939) with Sauer's genetic approach. This reconciliation took shape in programmatic essays, his own research on Prince Edward Island, where he combined cross-sectional studies with longitudinal analysis, and in his influence on Hartshorne's *Perspective on the Nature of Geography* (1959). As editor, he persuaded Hartshorne to include historical geography, now fixed in the profession by the work of Darby in Britain and Sauer in Latin America, as one of the mainstreams of the discipline.

Upon arrival in Wisconsin, Clark had hoped to take up once again his work on the overseas settlement of the British, but his first major project was a chapter on historical geography in *American Geography: Inventory and Prospect* (1954). This statement, which remained standard for at least a decade, laid down the principles of historical geography as the genetic study of 'areal associations and differentiations', and as thus distinct from history, which is concerned with 'human society in its various facets'. In this way he tried to preserve the disciplinary boundaries that Hartshorne had so painstakingly drawn, while at the same time allowing the legitimacy of the long view taken by Sauer and the Berkeley School. And, in common with both Sauer and Hartshorne, he held that historical geography is not a search for causes, since 'cause and effect relations are elusive', but is, rather, based upon the idea that geographical reality at any period is a momentary state in an indefinite continuum of change. For Clark, like Darby, historical geography was devoted to the recovery of past geographies, seen as points moving on a line of continuous regional transformation.

Clark held to this view for the remaining two decades of his life, reiterating it in a variety of programmatic statements and addresses. It decisively shaped his substantive work as well, which, apart from his contribution to the symposium on 'Man's Role in Changing the Face of the Earth', was all conceived in strictly regional terms. Returning in earnest to his research on the Maritimes, Clark began to organize materials for his books on Prince Edward Island and Nova Scotia. The first was originally conceived as a kind of methodological exercise, in which the relatively uniform physical geography and climate of the island, and the clearly localized settlement of different ethnic groups upon it, allowed Clark to isolate the effects of culture on economic and social patterns in a way that was not possible on the larger peninsula with its greater physical diversity and scattered ethnicity. The resulting volume, *Three Centuries and the Island* (1959), described the changing demography and economy of the island through an exhaustive series of maps. Grouped chronologically, these maps made up what Clark called 'instantaneous cross-sections', which he tried to link into a coherent, historical whole. In the event, however, this intention, as well as his goal of specifying the role of culture more concretely, was lost under the weight of the book's 155 maps and tables. Apart from the discovery of some revealing indices of the island's rural life, most notably the ratio of swine to sheep on individual holdings, the book failed to capture his own deep understanding of the island's distinctive regional experience.

Clark's last major work, and the most impressive testimony to his devotion to the Maritimes and regional reconstruction, was *Acadia: The Geography of Early Nova Scotia to 1760* (1968). At once more narrative and more comprehensive than its predecessor, this book meticulously traced the history of land tenure, demography, use of resources, and political shifts on the peninsula. Here Clark displayed his predilection for thoroughness to the fullest. Virtually every natural endowment, agricultural practice, linkage and settlement was explored to the limits of the available sources in order to document the evolution of the region through successive stages of European occupation. At the time of his death Clark was working on a second volume on Nova Scotia, organized as a series of thematic essays, covering the period from 1760 to 1867.

This very thoroughness lends his major scholarship a weight and bulk that tend to cast a shadow over his other work. During the last dozen-odd years of his life Clark also devoted considerable energy to the promotion of the interpretive element in geography and suggested ways of fostering it in students. These pieces reveal a more relaxed and discursive Andrew Clark, one with an unshakable faith in the mission of geographers to know regions and to bring them to life as dynamic, evolving wholes. He believed that geography in this comprehensive sense was indispensable and had, with history, a crucial function as 'world knowledge'. Toward the end of his life Clark gave substance to these humanistic convictions in several essays devoted to the problems of 'Canadianism'. Drawing on his roots and on his belief in the regional project, he took up the question of Canadian identity, the significance of the landscape, and the complexity of Canadian feelings about the United States. Clark's last publication might fittingly serve as an emblem of this reflective side of his work. 'The Look of Canada', originally an address given at an exhibition of Canadian landscape art at the University of Wisconsin–Madison, brings the main features of Canadian geography together in a broad sweep, from the soul of the Laurentian Shield to empires of resource extraction, in order to evoke the essence of the landscape and its bearing upon the people.

2. Scientific Ideas and Geographical Thought

Through programmatic statements, the training of new scholars, and by the example of his own research, Andrew Clark secured a place for historical geography as a distinct subfield in American academic geography. His antecedents were few. The only historical geographer of North America of any consequence in the years preceding Clark was Ralph H. Brown, but his was an isolated and idiosyncratic voice, and Clark, though mindful of his example, had no professional contact with Brown before his death. The major formative influences on Clark were the economic history of Harold Innis and the cultural-historical geography of Carl Sauer. Each was strongly orientated to the field and each pointed to an area of enquiry where nature and history come together: Innis, with his explanation of Canadian development powerfully situated in the physical context of the country, emphasizing its severity, its resources, and its routes; and Sauer, with his fertile conception of the natural setting under the impact of occupying groups. Clark absorbed and adapted their teachings to fashion his own object of investigation: the region as an ever-changing blend of nature and culture, the site of successive 'geographies of the past'. For this idea he perhaps owed more to the British historical geographer Clifford Darby, whom he first met in 1952 when Darby was travelling through North America, than to either Innis or Sauer. This trans-Atlantic connection was more firmly established when Clark spent 1954 as a visiting professor at University College London, and subsequently several of Darby's students spent a year working with him at Wisconsin. Darby's *An Historical Geography of England before A.D. 1800* (1938) and his Domesday Geography Series, which began appearing in the early 1950s, gave Clark powerful examples of the possibilities for reconstructing landscapes through the documentary record. In this way the region became the arena for his study of the modification of European ways overseas.

This emphasis on the region set Clark's basic approach apart from that of his mentors. Where Innis aimed at sweeping historical interpretations in his work, Clark felt that the reconstruction of past geographies 'must be done in terms of themselves, what they are and were and why and how they have come to be. In such terms we have an historical geography. The study of . . . the significance of these geographies to . . . human society is quite a different thing.' And Sauer eschewed a regional focus for geography altogether, preferring to focus on problems rather than on the characterization of places and areas. Clark, on the other hand, followed the neo-Kantian line of Richard Hartshorne in *The Nature of Geography*. Hartshorne argued that the fields of geography and history have no subject matter peculiar to themselves, but instead organize the plenum of reality along certain dimensions: for history this dimension is time, and for geography it is space. The region was therefore an indispensable framework in geography, and Clark believed that 'detailed and comprehensive knowledge of places . . . remains our basic *raison d'être*.'

This knowledge, he taught, is knowledge of patterns. And because these patterns are never fixed, but replace one another as frames in a 'steadily unwinding, never ending, cinematographic reel', Clark held that all geography is in an important sense historical: 'If the geographer is fundamentally interested in pattern he must therefore be interested in the changes in patterns, and to be a successful geographer he must concern himself with past geographies and with changing geographies through time.' These ideas, set down in the early 1950s, reveal Clark's strong affinity with Clifford Darby in mapping patterns of sequent occupance in order to depict regions as wholes at different points in their development. Clark, like Darby, attempted to confront the geography of change as it was illuminated by a series of historical cross-sections. After the publication of *Three Centuries and the Island* Clark became dissatisfied with a cross-sectional approach and turned instead to what he called 'geographical change', or, with a slightly different emphasis, 'the geography of change'.

These phrases express the essence of his mature teaching that the special interest and competence of the historical geographer lay in the changing patterns and associations of things in regions. Only by studying regions as 'continually changing entities', he believed, can we understand how present geography is rooted in the geography of the past: 'One cannot extrapolate from the point.' This principle guided his work on Maritime Canada. He conceived each of the areas he studied as a distinct geographical entity created by Old World settlers in the novel environmental, economic, and legal conditions of the New, and which evolved toward present patterns of occupance as those conditions changed. Accordingly, this work, and the larger enquiry into European settlement overseas of which it was a part, was not an antiquarian re-creation of the past, but an essay in self-understanding.

Toward the end of his career Clark sometimes referred to the theme of his overall research programme as 'The Suburbs of Europe' in order to reveal his scepticism about deterministic theories of cultural adaptation to the environment. For Clark, the most striking feature about our cultural geography was how strongly it still bore the imprint of its European antecedents after more than 300 years in the New World. As he put it in a prospectus written in 1953, he aimed to trace the persistence as well as the modification of this imprint in 'the development of patterns and practices of rural settlement and land use in the countries overseas settled from the shores of the North Sea (especially from the British Isles) in the 17th, 18th, and 19th centuries'. At this time he was projecting a series of regional monographs, with his book on New

Zealand as a first instalment, that was to have included work on the Maritimes, Australia, the eastern seaboard of the USA, South Africa, and some higher latitude areas of South America. Through comparison of the emerging of rural economies of these areas, Clark hoped to arrive at a general understanding of how nature and culture were blended in the overseas expansion of Europe. In the end, however, Clark himself was able to make relatively few comparisons or generalizations, completing only part of his work on the Maritimes and leaving the other areas almost entirely untouched. Nevertheless, the work of his students and of other scholars for whom he was a guide did document in other regions the patterns and processes of European overseas expansion.

Obviously, his personal research programme was overly ambitious from the start, but part of the reason for its lack of fulfilment lies in Clark's empiricism. Painstaking in the exhaustion of archival and field sources and meticulous in the reconstruction and mapping of the physical context, demographic movements, use of resources, and agricultural practices, he was uncomfortable with generalization, and he shied away from comparisons even of areas that he did know intimately. When he reviewed the theories of Harold Innis and Frederick Jackson Turner at the conclusion of *Acadia*, Clark found that they provided few, if any, insights into the particularities of Acadian history, and as for building his own interpretation through comparison, he wrote that 'scholarly prudence does not allow broad generalizations from the few such studies yet undertaken in North America.' It is characteristic that the three books he completed all deal with small, relatively isolated areas, two of them islands and one a peninsula, where he could at least approach the comprehensiveness that he sought. Even in Maritime Canada, the area he knew better than any other, he was unsure, writing two or three years before his death that 'from the beginning I felt that the whole area was unmanageable – and that even to hope to understand Nova Scotia, was, perhaps, rather optimistic.' He foresaw publication of his books on the Maritimes within a few years of his 1953 prospectus, to be followed by research and writing on Australia and South Africa. In fact the book on Prince Edward Island was delayed until 1959, and his work on Nova Scotia was split into two volumes, one published in 1968 and the other unfinished at his death.

At a time when theory-building and hypothesis-testing were in vogue in the discipline, it is not surprising that Clark was seen by some to be taking refuge from science in mere regionalism. In a review of *Acadia*, William Koelsch accused Clark of having no clear themes, writing 'textbookish' chapters, and amassing data unselectively; these charges were repeated, with less stridency, by Hugh Prince in an overview of historical geography a few years later. Clark responded by arguing that what is missing from a great deal of problem-orientated work in geography, and what regional geography did provide, is a *context* for generalization: in abstracting particular problems for study the general picture from which those problems were drawn is lost. For Clark, this contextual picture was essential, and 'almost indistinguishable from . . . that rounded view of changing regional character . . . which I see as the ultimate purpose of a very large part of historical geography'. In his own work on Acadia, for example, he claimed that the lack of a pre-existing exposition of the general historical/regional context meant that 'a rather full picture seemed . . . necessary. I couldn't discuss the variety of problems or put the generalizations in context if the context wasn't there.' The obvious target of his riposte was so-called positivist economic geography, but Clark also chided his cultural and historical colleagues in the Berkeley School for treating the past as a dead entity in which to seek the origins of traits or limited complexes of traits, and for taking on ecological questions without 'even a good description, let alone interpretation, of the general regional historical and geographical context in which they are set'.

In the end, however, Clark's commitment to regionalism was less an epistemological position than the response of a man who became a geographer out of his love for certain landscapes and the pleasure he got from studying them. Unlike Carl Sauer, he did not turn to rural ways and isolated areas as a conservative refuge from modernity but because he believed that their virtues enabled us to cope with the social and economic changes of modernization. It is hard to avoid the impression that Nova Scotia, where Gaelic was still a living tongue at mid-century, and Prince Edward Island, where the term 'foreigner' was applied to anyone from the mainland of Canada, as Clark reported with obvious relish, offered him an opportunity to celebrate and document the persistence of regional cultural diversity. 'The great forces afoot in the world,' he lamented, 'are striving to make us all more and more alike. I listen to some of my most liberal, decent, forward-looking friends at the University of Wisconsin talk of their plans for aiding the backward people of the world, backward as they view it. The end would seem to be a world of three or four billion Protestant, Republican, Rotarian, English-speaking, Chevrolet-driving, Midwest Americans . . . I find it a vision of hell on wheels and I want no part of it.'

Clark valued not tradition but diversity in itself, and also because he believed, by analogy with biology, that cultural vigour and innovation depend on variety, diffusion and mixing. He was profoundly discouraged by the self-ratifying vision of modernization that bleeds this vitality out of the world, abetted, he noted, by 'geographers who decide to devote themselves to planning, to "applied geography" in the service of the further and more rapid development of the affluent society'. Declaring himself instead with those 'who would lead the fight for the *survival of culture individualities*', Clark valued regions as distinct entities and devoted his considerable energy to fostering a different and more humane vision of modernization.

3. Influence and Spread of Ideas

At Wisconsin Andrew Clark occupied a place in academic geography that gave him a great deal of influence over the development of the discipline. During the formative period of his career, from 1940 to 1960, he found himself cutting across the two major currents of human geography, the culture history of Sauer and the regional, political, and economic studies advocated by Hartshorne as 'areal differentiation' and exemplified by the work of Vernor Finch and Glenn Trewartha at Wisconsin. Unlike the Berkeley School, which took the evolution of landscapes as its primary object and forged strong connections with anthropology through its concern with material culture traits, the morphology of landscape was a secondary interest to Clark. Socio-economic patterns and their transformation were more basic to his approach, and this focus led him to history rather than anthropology for issues and themes. Fieldwork, which for Sauer and his followers nearly achieved the status of a cult, was always overshadowed in Clark's work by his meticulous handling of primary documents. What was for the Berkeley School 'cultural' historical geography became for Clark 'historical' historical geography. His alliance with historiography may have also distanced Clark from the disciplinary limits implied in *The Nature of Geography*. Though he shared with Hartshorne a strong regional orientation, his focus on the process of regional change took him beyond simply charting distributions and differences over areas, and his method of carefully interrogating his sources set him apart from the superficial mapping and cataloguing of traditional regional geographers.

Whatever differences Clark may have had with other geographers, the mastery he displayed over the documentary record and his circumspection in interpreting it earned him a reputation as a rigorous and scrupulous scholar. As a result, he was given the task of editing the Association of American Geographers' monograph series, the profession's first high-profile venture designed specifically to present a respectable face to other academic groups. Among these monographs was a study of the South Australian Wheat Frontier by Donald W. Meinig, and in this editorial role, he developed an enduring respect for the other dominant figure of North American regional historical geography. He also conceived and edited a series of books on the historical geography of North America for Oxford University Press, which also included another major work of Donald Meinig on the American Southwest. In his final years he helped to launch *The Journal of Historical Geography* as its first co-editor for North America.

Throughout the 1960s Clark's influence in the subfield of historical geography was enormous, but by the end of that decade he was concerned about the marginality of historical geography as human geography itself was transformed. The earlier debates between Sauer and Hartshorne were replaced by issues derived from the new social scientific methods of modelling, quantification and hypothesis testing. Clark opposed these 'gurus of the new wave', but he was reacting as much to their ideological style as to their substance. The fog of obscure and barbarous language blowing off 'the gobbledygook boys' disturbed Clark. Though he once worried to a correspondent that his sentences were 'long, involved, Teutonic monstrosities', his prose was generally clear and straightforward; if it lacked pungency, it was always free of 'jargon, pretension, and the assumption of an esoteric professional persona', which he believed were usually 'smokescreens thrown up by second-class minds to hide their own inadequacies'.

Clark's background in mathematics and statistics gave him an understanding of the limitations of quantification in geography, but he was not afraid of it. On the contrary, in the 1960s he once proposed computerizing various census data from the nineteenth century and displaying them on a large glass-fronted map with coloured lights of variable intensity keyed into hexagonal cells. Within the limits of existing technology he was describing cartographical techniques that are now commonplace. His aim, as always, was better to portray geographical change, and also to begin to map rates of rate of change in significant indices, or 'the locational pattern of change itself'. This is what he meant by the term 'geography of change'.

The drive toward purely spatial generalizations was alien to Clark because he saw the fruitful possibilities of applying geographical perspectives to historiographical questions. In his own work this connection was not explicit, apart from the discussion of the metropolitanism of Harold Innis and Frederick Jackson Turner's frontier thesis at the end of *Acadia*. His books were not arguments in the logical sense, but representations. Of *Acadia* he wrote that 'with its concern to read out of the Acadian experience whatever may seem of value in reflection on general social and economic theory, the primary purpose of the book . . . is to describe the changing character of Acadia and the Acadians, as a geographer views them retrospectively.'

Historiographical relevance was, however, a defining feature of many of the dissertations Clark supervised. Paul W. English, Clarissa T. Kimber and Robert Hofpauir worked in a traditional Berkeley mode, and David Ward extended historical geography into the realm of the industrial urban life of the nineteenth century. The remaining dissertations were closer to his own research and fell into two groups: firstly, the creation and transformation of various colonial frontiers of North America (James R. Gibson, R. Cole Harris, Grand C. Head, James T. Lemon,

H. Roy Merrens, Phillip A. Muntz and Arthur J. Ray); secondly, the settlement and economic development of the American frontier of the nineteenth century (Leonard W. Brinkman, Sam B. Hilliard, Terry G. Jordan, Thomas F. McIlwraith, Aidan McQuillan, Stephen L. Stover and Ralph D. Vicero). These studies addressed specific historical issues such as the mediation of ethnic cultures in frontier society, subsistence vs. commercialism, and core/periphery relations. Clark encouraged his students to delve into political and economic matters that were not, at the time, seen to be part of the geographer's purview. This aspect of his teaching achieved clearest expression in a paper he delivered to the plenary session of the sixtieth annual meeting of the Association of American Geographers in 1964, 'The Coordinates of Historical Geography'. In this address Clark put out a sweeping call 'to relate distributional changes to social and economic theory', and he admonished historical geographers to 'stop pretending that the industrial and urban revolution hasn't occurred'. Despite this commitment to major historiographical issues, it would be an exaggeration to describe this agenda as a 'school' for in many respects the genius of Clark as a teacher was that he did not emasculate his students, particularly once they had established their independent professional careers.

In the heyday of those who sought to legitimize geography as an independent science of space, Clark was pushing his students and his colleagues instead to apply their special interests and abilities as geographers to question generalizations that were not bounded by any one discipline, a position that was to become standard twenty years later. Clark was remarkable for his generation in the degree to which he and his students contributed to the changing perspectives of American historians. This breach of disciplinary autonomy offended his principal antagonists, the ideologues of spatial analysis, who were committed to an increasingly narrow view of human geography. Other historical geographers, particularly those working in the shadow of Carl Sauer and who increasingly described themselves as cultural geographers, questioned the need for an alternative historical geography and questioned his indifference to morphological studies. Perhaps in response, and perhaps out of dissatisfaction with the scholastic empiricism of *Acadia*, in his last years Clark turned increasingly toward the humanistic appreciation of landscapes. His essays on Canada and Canadianism and his general statement on the art of knowing and describing regions, 'The Whole Is Greater than the Sum of Its Parts', anticipate, in part, the redefinition of cultural geography round the social production of landscapes that came in the decade and a half following his death. Andrew Clark never resolved the antinomy of his relentless pursuit of the facts of regional change and his desire for a respectful and nostalgic evocation of landscapes; but each impulse, held suspended and exaggerated in his geographical make-up, looked ahead to a more expansive and plural human geography in which he might have felt more at home.

Bibliography and Sources

1. OBITUARIES AND REFERENCES ON ANDREW HILL CLARK

Harris, R. Cole, 'Andrew Hill Clark, 1911–1975', *J. Hist. Geogr.*, Vol. 2 (1976), 1–2.

Koelsch, William A., Review of *Acadia, Econ. Geogr.*, Vol. 46 (1970), 201–2.

Meinig, D.W., 'Andrew Hill Clark, historical geographer', in James R. Gibson (ed.), *European Settlement and Development in North America: Essays on Geographical*

Change in Honour and Memory of Andrew Hill Clark, Toronto, University of Toronto Press, 1978, 3–26.

Prince, Hugh C., 'Real, imagined and abstract worlds of the past', *Prog. Geogr.*, Vol. 3 (1971), 4–86.

Ward, David, 'Andrew Hill Clark, 1911–1975', *Ann. Assoc. Am. Geogr.*, Vol. 67 (1977), 145–8.

Warkentin, John, 'Epilogue', in James R. Gibson (ed.), *European Settlement and Development in North America*, pp. 208–20.

2. *PRINCIPAL PUBLISHED WORKS OF ANDREW HILL CLARK*

a. Books

1949 *The Invasion of New Zealand by People, Plants and Animals: The South Island*, New Brunswick, NJ, Rutgers University Press, 465 pp.

1959 *Three Centuries and the Island: A Historical Geography of Settlement and Agriculture in Prince Edward Island, Canada*, Toronto, University of Toronto Press, 287 pp.

1968 *Acadia: the Geography of Early Nova Scotia to 1760*, Madison, University of Wisconsin Press, 450 pp.

b. Major Articles in Historical Geography

1945 'The historical explanation of land use in New Zealand', *J. Econ. Hist.*, Vol. 5, 215–30.

1947 'South Island, New Zealand and Prince Edward Island, Canada: A study of insularity', *New Zealand Geographer*, Vol. 3, 137–50.

1950 'Contributions to the geographical knowledge of Canada since 1945', *Geogr. Rev.*, Vol. 40, 280–308.

1954 'Titus Smith Junior and the geography of Nova Scotia in 1801 and 1802', *Ann. Assoc. Am. Geogr.*, Vol. 44, 291–314.

1956 'Canada and Australia: A comparison', *Geogr. Rev.*, Vol. 46, 421–23.

1956 'The impact of exotic invasion on the remaining New World mid-latitude grasslands', in William L. Thomas (ed.), *Man's Role in Changing the Face of the Earth*, Chicago, University of Chicago Press, 737–62.

1960 'Old World origins and religious adherence in Nova Scotia', *Geogr. Rev.*, Vol. 50, 317–44.

1962 'The sheep/swine ratio as a guide to a century of change in the livestock geography of Nova Scotia', *Econ. Geogr.*, Vol. 38, 38–55.

1965 'New England's role in the underdevelopment of Cape Breton Island during the French regime, 1713–1758', *Can. Geogr.*, Vol. 9, 1–12.

1975 'The conceptions of the empires of the St. Lawrence and the Mississippi', *Am. Rev. Can. Stud.*, Vol. 5, 4–27.

c. Geography as a Field of Study

1954 'Historical geography', in P.E. James and C.F. Jones (eds), *American Geography: Inventory and Prospect*, Syracuse, NY, University of Syracuse Press, 70–105.

1960 'Geographical change – A theme for economic history', *J. Econ. Hist.*, Vol. 20, 607–16.

1962 'Praemia Geographiae: The incidental rewards of a professional career', *Ann. Assoc. Am. Geogr.*, Vol. 52, 229–41.

1972 'Historical geography in North America', in A.R. Baker (ed.), *Progress in Historical Geography*, Newton Abbot, Devon, David and Charles, 129–43.

3. PRINCIPAL UNPUBLISHED ADDRESSES

1947 'Legend and fact in historical geography: An illustration from Nova Scotia', forty-fourth annual meeting of Assoc. Am. Geogr., Charlottesville, VA.

1964 'Coordinates of historical geography', plenary session of the sixtieth annual meeting of Assoc. Am. Geogr., Syracuse, NY.

1966 'The three worlds of the Canadian', plenary session of the sixty-second annual meeting of Assoc. Am. Geogr., Toronto.

1974 'The whole is greater than the sum of its parts: A humanistic element in human geography', Symposium on Changing World-Views: A Half-Century of Geography at Michigan, Ann Arbor, MI.

David Ward is Andrew Hill Clark Professor of Geography and Provost and Vice Chancellor for academic affairs at the University of Wisconsin, Madison, USA. Michael Solot is a doctoral student at the same institution.

Chronology

1911	Born on 29 April in Fairford, Manitoba
1926	Graduated from Brandon Collegiate Institute
1926–31	Summer surveyor's assistant and labourer for Manitoba Power Commission and Manitoba Good Roads Board
1930	BA in mathematics and physics, McMaster University
1931–32	Postgraduate courses in statistics, University of Manitoba
1932–35	Actuarial assistant, Manufacturer's Life Insurance Co., Toronto
1935–38	Assistant and demonstrator, Department of Geography, University of Toronto
1936–37	Field assistant, Dominion (of Canada) Geological Survey
1938	MA by examination in geography, geology, and economic history, University of Toronto; summer field reconnaissance of western Europe and north Africa with Griffith Taylor

1938–40	Teaching assistant, University of California, Berkeley
1939	Summer fieldwork on Prince Edward Island
1941–42	Lecturer in meteorology, US Army Air Force, Berkeley
1943–44	Programme lecturer in regional geography, US Army
1944	Ph. D., University of California, Berkeley; naturalized US citizen
1944–46	Geographer, US office of Strategic Services and Department of State, Washington, DC, and China
1946–49	Associate professor of geography, Rutgers University
1949	Publication of *The Invasion of New Zealand*; named chairman of first department of geography at Rutgers
1949–51	Professor of geography, Rutgers University
1951–75	Professor of geography, University of Wisconsin–Madison
1954	Presented lectures in geography at University College London in May; archival work on British in North America
1957–61	Chairman of department of geography at UW–Madison
1959	Publication of *Three Centuries and the Island*
1961	Honorary President of the Association of American Geographers; named Brittingham Research Professor, University of Wisconsin
1961–62	Fulbright Research Scholarship, UK
1962	John Simon Guggenheim Memorial Fellowship
1966	Named Vernor Clifford Finch Research Professor of Geography, University of Wisconsin–Madison
1968	Publication of *Acadia*
1969	American Historical Association Beveridge Citation of *Acadia* as best historical work on Canada for 1968
1971–75	Editor, Oxford University Press series in the Historical Geography of North America
1971–72	Fellow in American Studies and visiting professor at the University of Dundee, Scotland
1974	Canadian Association of Geographers Award for Scholarly Distinction in Geography; visiting professor (May and June), Hebrew University of Jerusalem
1975	Co-editor of *Journal of Historical Geography*; honorary fellowship in American Geographical Society
1975–76	Named Claude Bissell Professor of Canadian and American Studies, University of Toronto (unfulfilled)
1975	Died 21 May in Madison, Wisconsin
1975	D.Sc., *honoris causa*, Brandon College, Manitoba (posthumous)

Clarence James Glacken

1909–1989

Anne Macpherson

Clarence J. Glacken began his career in academic geography late in life, in 1952, at the University of California, Berkeley, where he remained until his retirement in 1976. His greatest contribution to the field of geography is his view of the significance of ideas about the relation of man to the physical environment as expressed on landscapes and in the direction of study. He recognized three dominant themes in the history of thought that have persisted from antiquity to the present. His work is summed up in his great classic study, *Traces on the Rhodian Shore*, published first in 1967, in which he describes the permutations of these three leading ideas through the eighteenth century. The volume has been in steady demand since its publication.

1. Education, Life and Work

Clarence J. Glacken was born on 30 March 1909, in Sacramento, the capital of California. His family had deep roots in the city – his maternal grandmother born there, his paternal grandmother brought there across the plains as an infant, and both grandfathers settling there as young men. His fond memories of boyhood were deeply imbued with his sense of the place, at the confluence of the American and Sacramento rivers, terminus of the Pony Express and cross-continental railway, and a hub of activity during the Gold Rush in 1849. He used to visit Fort Sutter in the city and Coloma, where gold was discovered, and take the ferry through the tules of the Delta to San Francisco. In those days Sacramento was smaller, and much was accessible to a boy on his bike. He could have the thrill of exploration, of watching the waters of great rivers mingle, of crossing boundaries – albeit county – without needing to travel far. Indeed, by the time he was twenty-one he had not even been to Los Angeles, but 400 miles was a great distance in the days of the Model T. One feels that this sense of place, of belonging, had much to do with what interested Clarence Glacken, the geographer, in later years.

Aside from exploration and his appreciation of the locale, Glacken mentions other boyhood interests that could be termed geographical (although he says his first ambition was to be a streetcar conductor). As a boy he started a stamp collection and was given a huge album by his grandfather when he was fifteen. The stamps from countries and colonies with wonderful names, and the beautiful scenes they often showed, stimulated his interest in far-off places. One of his primary school teachers was interested in geography and Glacken enjoyed the geography bees, when students would have to find places on the map. He early developed an interest in Asia and the Far East.

Glacken speaks favourably of his high school and junior college teachers, who were 'encouragers, gentle but firm correctors of incorrect English' (Buttimer, p. 22). In junior college he read Isaiah Bowman's *The New World: Problems in Political Geography*, a study of the world after World War I and the Treaty of Versailles, and he was especially impressed by the wealth and variety of the maps. A young English instructor at Sacramento Junior College, John Harold Swan, suggested that Glacken take a course from Teggart when he transferred to Berkeley. Teggart's course introduced Glacken to the history of ideas.

Frederick J. Teggart was a power unto himself at the University of California. A dynamic teacher, he was also a force in campus politics. Although he worked with Carl Sauer, Alfred Kroeber, Paul Taylor and others to improve support for social science research at the university and nationally, he was hard to get along with and was relegated to his own department, the Department of Social Institutions, later to become the Department of Sociology.

When Glacken entered the University of California as a junior in 1928, he took Teggart's year-long course, 'The Idea of Progress'. More than a social history, the course was a history of history, of ideas and social theories. Progress, in Teggart's view, was a comparatively new idea in Western thought. His course and the readings he collected awakened Glacken's understanding of different ways of thinking in the past and in different cultures, and of the importance of the past. Teggart's course did more than introduce Glacken to a new field of study, intellectual history. It was, he later claimed, an 'exposure to excellence' (Pred interview). He was so excited by Teggart that he switched his major from political science and received both his BA and his MA from the Department of Social Institutions.

Margaret Hodgen was an assistant professor in the Department of Social Institutions, interested in the history of cultural diffusion and social change, the history of social thought. Her course introduced Glacken to environmental determinism and other geographical ideas and to geographers such as Ratzel and the French possibilists. Although he had heard of Sauer, he was too taken up with Teggart and intellectual history to take any courses from him, or from Kroeber, Lowie or other eminent scholars on campus. Glacken and Hodgen became friends, exchanging ideas about their work and suggestions for stimulating reading until her death in 1977.

Glacken received his BA in 1930 with the highest honours. He also studied German and continued his interest in the Far East by taking a course in Far Eastern politics and history from a visiting Chinese professor. Some of the books he read at that time made a lasting impression. Teggart's *Prolegomena of History* and *Theory of History* are fundamental analyses of history and historiography. Edgar Arthur Burtt's *The Metaphysical Foundations of Modern Physical Science* is an investigation of the metaphysical ideas, assumptions and religious beliefs behind the discoveries of the founders of modern science. John Livingston Lowes's *The Road to Xanadu* became a special favourite. This study of the sources of Coleridge's imagery in *The Ancient Mariner* and *Kubla Khan* and the process of transforming images into poetry, and his own later rereading of Coleridge's poetry, convinced Glacken that Coleridge was the most ardent student of geography among English poets (Buttimer, p. 23) and that aesthetic ideas and the study of art and literature have a place in geography.

Glacken received his MA in Social Institutions by examination in 1931, at the outset of the Depression. His fields of study were history and methodology of the social sciences, theory of history, history of the idea of progress, and historical geography of Eurasia in late Roman times. Fellowships had been rare at best and at that time were non-existent because of the Depression. Glacken worked as a research assistant in the Department of Social Institutions after graduation, studying original source materials of the late-Roman period in the West, and made a survey of the causes advanced for the fall of the Roman Empire (climatic, economic, psychological, etc.). After a couple of years getting along on odd jobs and tutoring, unable to pursue his academic studies, he started his career as a bureaucrat.

After the inauguration of President Franklin Roosevelt, many federal and state relief agencies were established. In 1934 Glacken became a case worker and county director in northern California for the State Emergency Relief Administration. In 1936–37 he was a research assistant for the Resettlement (later Farm Security) Administration, surveying possible sites for migrant labour camps. When he was not travelling to various places in northern California, he lived in San Francisco. Exploring 'Little Italy' and Chinatown and the working Fisherman's Wharf intensified his earlier interest in cities. From 1938 to 1941, after returning from an eleven-month journey round the world, he became a supervisor of relief work, first in Oakland and then in Tulare, where he supervised preparation of a rural housing survey.

His work for these relief agencies brought Glacken into close acquaintance with the migrant 'Okies' and 'Arkies' so movingly described by John Steinbeck in *The Grapes of Wrath*. Their stories made him aware of the importance of different forms of land tenure in man's relation to the land. In this time of the Dust Bowl he first became interested in soil and soil conservation, a continuing concern which, along with the relation of resource use to poverty in general, prompted his eventual decision to enter postgraduate study in geography at Johns Hopkins University.

From February to December 1937 he realized a dream he had had since his teens – to travel. He had saved money, the exchange rate was favourable, and tourist and third-class fares on ocean liners were remarkably cheap. He saw the world from train and ocean liner, not 'the world of airports which even now I find strange and disembodied' (Buttimer, p. 24). In his 1952 biographical sketch, Glacken states that the purposes of the trip were to find out about geography, culture, population problems, standards of living, and conservation of natural resources; to visit sites of some ancient civilizations; and to see parts of Europe and Asia before the war that was obviously coming. He later declares, 'I did not go as an accomplished trained observer, but as an ordinary young traveler with an AB. My experiences intensified what my studies had suggested; the profound importance of cultural differences. . . . In retrospect, I look upon my travels as a species of field work. . . . In the light of subsequent interests, this was field work in an entirely different sense, a preparatory experience.' He thought he would not have developed his intense enthusiasm for the history of ideas without the vivid experiences of that trip. 'It would have been a world of abstractions. . . . I have a mind that needs the concrete. . . . When I read about ideas and their history and they are related to places I know, both become more vivid, full of meaning' (ibid., p. 25). He goes on, reflecting on the later painful conflicts in geography: 'Travels like mine are powerful reminders of the limitations of theory, of the pitfalls of models, of what in the 1960s became by some a virtual apotheosis of nomothetic thinking.'

His itinerary took him from Yokohama to Moji in Japan; to Tientsin, Peking, Nanking, Shanghai, Canton and Hainan in China; Hong Kong; French Indo-China, including the ruins of Ankor Vat; Singapore; Colombo; Bombay; Massaua in Eritrea; Cairo and Alexandria in Egypt; through Palestine and Syria; cycling round Cyprus; Athens, Knossos on Crete and by ship to Istanbul; on an Intourist trip through the USSR; Finland, Sweden, Denmark, Germany, Austria, Italy, France and the UK. 'All this has intensified earlier as well as subsequent feelings in me, a respect for individual experience, the discrete, the concrete, and a critical attitude toward theory, especially Western theories arrogantly striving to be ecumenical' (ibid., pp. 25–6).

Glacken mingled as much as he could with the people of the countries he visited, travelling third class on trains and walking the streets of cities, and was often invited into homes. These experiences, reinforced by later ones as administrator in Korea during the Occupation and his field study on Okinawa, freed him from 'parochialism based on experience and study of Western civilization alone' (ibid., p. 30). He was distressed by simple-minded contrasting of East and West as though the differences between China and India were not as great as those between either of them and Western civilization.

Returning from his *Wanderjahr*, Glacken again became a bureaucrat, first in Oakland for the WPA, then in Tulare for the State Relief Administration, supervising unemployment relief and the preparation of a rural housing survey. Three months after his first wife died in 1941, before the Japanese attack on Pearl Harbor, he was drafted into the US Army, where he served for six years. After a period as an enlisted man, he was sent to Officer Candidate School at Ft Washington, Maryland. He applied for Civil Affairs Training School and Japanese language study and was accepted, probably because of his travels in the Far East and the courses he had taken at the University of California.

During 1943 and 1944 at the University of Chicago he studied intensively Japanese culture, history, economics, politics and for three hours a day, six days a week, the Japanese language, written and spoken. *Suye Mura*, a study of village life on Kyushu by his instructor in Japanese culture, the anthropologist John Embree, was a great help to Glacken in his later studies of village life on Okinawa. On completion of the course, he taught military government at the Civil Affairs Training School on the Chicago campus of Northwestern University. From the Occupation in 1945 until June 1946 he served in the Military Government as Deputy Director of Health and Welfare, stationed in Seoul. He liked the old parts of the city and visited Kaesong, Taejon, Taegu and Pusan. He was impressed by the extent of deforestation in Korea.

Released from the Army, Glacken returned to Carlisle, Pennsylvania, to teach in the US Army School for Government in Occupied Areas, from August to December 1946. He lectured on Japanese and Far Eastern history, the economics of military occupation, customs, people, and aspects of military government in Korea. From 1946 to 1949 he worked as a research analyst and statistician in the Office of Vocational Rehabilitation and Education of the Veterans Administration in Washington, DC. In 1948–49 he was also a part-time lecturer in history at American University. He assisted in planning and taught a required course, 'American Democracy in Its World Setting'. The topics reflect his continuing and later interests: geography, the age of discovery and subsequent history of European expansion, economic development from the Middle Ages, the growth of political institutions and political thought, the outlines of religious history, the beginnings of modern science and the history of technology, and the quest for world order.

At this time, Glacken concluded that the only occupation in which he could find real satisfaction was that of university professor. He had had plenty of experience as a bureaucrat. In his personality there was not a trace of the competitiveness that would make a career in business attractive. By now he had remarried and had two children (Karen and Michael), but his wife, Mildred, supported him in his decision to change careers.

When Glacken was at Officer Candidate School, Mildred had taken a job in the office of the director of Inter-American Affairs as assistant to William Vogt, and Glacken met Vogt through her. They shared common interests and became friends. Vogt had studied environmental destruction in Latin America, and his book *Road to Survival* was an effort to make people, especially those in power, aware of the dangers of continued population growth and the resulting worldwide destruction of the environment. In 1949 the United Nations held a Scientific Conference on Conservation and Utilization of Resources, further spurring Glacken's interest in the relation of population and resources, in soil erosion and poverty, first aroused by his experiences in the San Joaquin Valley and his subsequent travels. He wanted to delve into the history of ideas on the subject. Vogt suggested that he attend the Isaiah Bowman School of Geography at Johns Hopkins University in Baltimore. There he received his Ph.D. in June 1951.

Glacken's main fields of study at the Isaiah Bowman School were human geography; history of geographical ideas; geography of soils; tropical life; physiological climatology; political geography; economic geography, including economic theory as applied to geographical problems; descriptive oceanography; and history of cartography. He wrote his dissertation, 'The Idea of the Habitable World', under George Carter and E.F. Penrose. It was never published, but as he continued to learn about the complexities of the periods and ideas, the dissertation evolved into his *Traces on the Rhodian Shore*.

The thesis of the dissertation was that the history of ideas can put into perspective contemporary controversies regarding population growth and its relation to the use and preservation of natural resources. He strove to point out the importance of ideas, from the eighteenth century to the present, which have dealt with the relationships between population, the cultural and economic activities of mankind, and the natural environment. The main themes of the dissertation were the relation of population growth to resources, the influence of the geographical environment on human culture, the contrasting idea of man as a geographical or geological agent, concepts of the soil, and living nature as a series of complex relationships. He investigated the writings of such authors as Malthus; Carl Ritter and Alexander von Humboldt; George Perkins Marsh, Jethro Tull, and Justus von Liebig; and Darwin, Huxley, and Wallace.

In 1951 the Pacific Science Board of the National Research Council undertook a series of studies of the way of life of the peoples of the Ryukyu islands. Through his association with the anthropologists who had directed the Japanese language programme at the University of Chicago, Glacken was invited to take part. He lived on Okinawa in 1951–52, studying the life of three villages, two agricultural and one fishing. He organized his study around three themes; the family system, a blending of the Chinese and Japanese systems; the land tenure system, the division of land and the effects of the inheritance system; and the actual results of these in the use and appearance of the land. He found the experience personally rewarding, and had a deep respect for the people he lived with. His study resulted in *The Great Loochoo*, published in 1956 by the University of California Press.

While Glacken was on Okinawa, Mildred and the children lived in Berkeley. On his way overseas, he called on Sauer at the University of California to pay his respects. He again called on Sauer after he had returned to Berkeley and was working on his report. On one visit, Sauer asked him, 'so casually that I did not know whether it was an offer or not', whether he would like to join the Department of Geography (Buttimer, p. 28). Only when he was assigned an office was he sure, and started his career at the University of California in the autumn of 1952. He was grateful for the opportunity to return to the university from which he had been graduated twenty-two years before. 'I had always felt like a Far Westerner, not in a chauvinistic sense, but the place I knew and felt most at home in was the Bay Region. . . . I felt I belonged there, needing no guides to the terrain' (ibid.).

Geography was then a small department housed in the basement of Giannini Hall. Glacken made the sixth member of the faculty, joining Sauer, John Leighly, J.E. Kesseli, James Parsons and Erhard Rostlund. The department under the chairmanship of Carl Sauer had developed a distinctive style, the Berkeley School of cultural geography, looking at landscapes created over time by peoples according to their different cultures, with an emphasis on rural areas. Because of proximity and the resources of Bancroft Library, as well as interest, Sauer and most of his students concentrated on Mexico and the American West. Glacken had a similar point of view toward geographical study, but with more emphasis on the non-material aspects of culture. His experience in the Orient and Europe and deep interest in the history of ideas nicely complemented the strengths of the rest of the faculty.

In his first five years at Berkeley, first as instructor and the second year as assistant professor, Glacken taught the introductory economic geography sequence; the Saturday field course; courses in the geography of the Far East, Northern Europe, and the Mediterranean; and Geography 176 – 'Relations between Nature and Culture', the new course he developed, taught regularly until he retired and was twice recalled to teach. He participated in the extraordinary 1955 international symposium, 'Man's Role in Changing the Face of the Earth', to which he contributed a paper, 'Changing Ideas of the Habitable World'. The two-volume publication of the symposium has gone through several printings and is still in print.

The Great Loochoo, his study of Okinawa, was published in 1956 by the University of California Press. While in Norway in 1957–58 as a Fulbright scholar studying problems of conservation, he was introduced to the *seters*, the upland pasture chalets where people moved in the summer to graze their cattle and make cheese, an extension of the ancient system of transhumance. Aesthetically he was enchanted with the variety of glacial landscapes, each seemingly 'stolen from an illustration in a textbook on geomorphology' (Buttimer, p. 31).

On his return from Norway, Glacken, now associate professor and departmental major adviser, took over the introductory course in cultural geography, initiated by Carl Sauer, and began teaching the seminar in that subject. Paul Wheatley, whose special interest was non-Western cities, had joined the faculty. His interests and Glacken's were complementary and provided mutual stimulation. They alternated in teaching the regional courses on Europe and the Far East and the seminars in cultural and historical geography. In 1962 they developed the two-semester, upper-division advanced course in cultural geography and taught it together until 1966. Glacken now taught American geographical thought and continued his nature and culture course, Geography 176, now described as 'A history of the great ideas in

Western thought from antiquity to the present, concerning the relationship of human culture to the natural environment'.

Continuing his investigation of Western intellectual history, Glacken made an intensive study of Count Buffon's *Histoire naturelle*, paying special attention to the descriptions of environment changed by human activity. 'Count Buffon on Cultural Changes of the Physical Environment' was published in the *Annals of the Association of American Geographers* in 1960. He contributed to another major symposium in 1961, 'Man's Place in the Island Ecosystem', organized by F.R. Fosberg for the Tenth Pacific Science Congress in Honolulu. His paper was 'This Growing Second World Within the World of Nature'. Meanwhile, he was organizing his ideas and beginning to write *Traces on the Rhodian Shore*. He was made full professor in 1964.

Glacken was awarded a $6000 Guggenheim Fellowship in 1965 to study the history of the nature protection movement. After his great book was published by the University of California Press in 1967, he left for Europe where he spent the autumn and winter quarters, visiting several nature protection organizations in the UK and Switzerland. He began to plan the sequel to *Traces on the Rhodian Shore*, which had carried the history of the leading ideas about man's relation to nature only through the eighteenth century. In 1968 he received the Association of American Geographers Citation for Meritorious Contributions to the Field of Geography 'for his monumental *Traces on the Rhodian Shore*, an examination of ideas about man and nature exemplary of the greatest scholarly tradition'.

The late 1960s was a time of political and social turmoil nationally and especially in Berkeley, of dissension in the field of geography, and of tension among the geography faculty at Berkeley. Glacken had been spared the bitterness of the loyalty oath controversy which had shaken the campus from 1949 until a sort of compromise was reached in 1952. (His friend Margaret Hodgen was a non-signer and had been suspended for some time.) During the 1950s and early 1960s he was able to contemplate the long history of ideas about man's relation to nature in relative calm and peace, but the Free Speech Movement and consequent postgraduate student strike shattered the quiet of the Geography Department in the autumn of 1964, and began the years of campus turbulence.

Other stresses developed on campus when the university converted to the quarter system in 1966, to the displeasure of many of the faculty. This change required re-examination of programmes and revision of courses to a greater or lesser degree. In the Geography Department, semester courses merely became quarter courses with some renumbering for the most part, but all was far from peaceful on the fifth floor of the Earth Sciences building.

In the late 1950s and early 1960s the field of geography was torn by controversy as the urban-economic, theoretical, and quantitative geographers were challenging the traditional practitioners of the discipline whose emphasis had been on areal studies and the uniqueness of place – the Quantifiers vs. the Qualifiers. This became more than a gentlemanly philosophical difference when university power politics, budget requests and staffing were at stake. Geography departments became vulnerable to attack by the 'hard' sciences. Faculties were expanding as student populations grew, and within departments the 'Young Turks' were jockeying for power. The Department of Geography at Berkeley was not spared these conflicts. The faculty was divided into factions either defending the 'Berkeley School' that had developed under Carl Sauer or trying to turn geography's course into a more 'modern' direction. Faculty rivalry was intense, affecting even the students.

Meanwhile the students, both postgraduate and undergraduate, were becoming involved and their basic faiths challenged by the political controversies rocking the campus, the city, and the nation, which carried such strong overtones of morality, of justice, and of right and wrong. Their concerns also divided the faculty. (For an account of the controversies and disturbances on campus, see W.J. Rorabaugh, *Berkeley at War: The 1960s*.) The student body had become older and from at least 1957 more radicalized, while the administration and Board of Regents and Ronald Reagan's state government were more conservative. Students and faculty were genuinely concerned about the Vietnam War, the draft, and the civil rights struggles in the South and closer to home, however much they decried the methods of protest. By the autumn of 1966 the radicals wanted action to gain power, and focused on universities to attract disaffected students. So began the pattern, as Glacken once observed, of a disturbance the fifth week of every quarter, usually bringing the police onto campus, incidents generally intensifying each quarter in violence and virulence.

In 1966 Glacken became chairman of the Geography Department, as senior faculty member and everyone's choice. As chairman, he faced the difficult task of keeping the department functioning. He returned from Europe in early 1968 to the sick atmosphere on campus. Roger Heyns, a moderate, had been appointed chancellor to try to bring order to the campus, but his efforts were unavailing, unsupported by the 'paranoic' Regents in their power struggle with the radicals, who deliberately created incidents to bring police and tear-gas on campus. Throughout 1968 the antiwar movement and draft resistance grew in intensity. The largely black Third World Liberation Front was formed to press for ethnic studies courses, mounting some particularly disturbing marches and incidents in 1968. There were arson fires in Wheeler Hall, the Main Library, and several other buildings during the next several years as the demonstrations and reactions increased in fury, culminating in May 1969 in the People's Park lunacy and six days of turmoil; occupation by the Highway Patrol, sheriff's deputies, and the National Guard; and tear-gas sprayed from a helicopter. Finally, in the spring of 1970, the US invasion of Cambodia, and the Kent State and Northern Mississippi State killings brought such an explosive and almost united reaction from the student body that classes were stopped, and the university was 'reconstituted', as students of each department organized to bring their outrage and protest against the war to the public at large.

These years were extremely trying for such a sensitive, gentle and conscientious man as Clarence Glacken. He was opposed to the war in Vietnam, had travelled through Indo-China on his round-the-world trip, had deep sympathy and respect for the people there, and had a draft-age son. At the same time, he felt the obligation of his position to teach his classes without apology for the ideas he had for so long considered important and still 'relevant', and to keep the Geography Department functioning as much as possible. He barely managed to hold on until the end of the spring quarter 1970, and then had a severe nervous breakdown. That autumn he had a massive physical breakdown and was six months recuperating. By the spring of 1971 he was able to return to campus to teach his seminar, 'Topics in Cultural Geography', and supervise his students' theses and dissertations.

He taught advanced cultural geography, geographical thought, his own course on man and nature, and seminars until 1976, except for a sabbatical in autumn and winter 1974–75. He retired in 1976, but was called back in spring 1979 and winter 1981 to teach his course. In 1976 he was awarded the Berkeley Citation from the University of California 'for distinguished achievement and for notable contributions to the University'.

Glacken continued to delve into the writers of the nineteenth and early twentieth centuries, gathering material for a sequel to *Traces on the Rhodian Shore*. He was finding an enormous amount of material, and his notions of which ideas needed attention were changing. He wrote to Margaret Hodgen in October 1974 that he had had a heart attack in March, and that he had recovered but could not work at the sustained pace he needed for work on the nineteenth century (Bancroft, Hodgen Papers). He had been looking into the area of subjective, aesthetic and emotional attitudes toward the natural world, and told David Hooson in his 1980 interview that he was almost finished with that section (*Geographers on Film*, July 1980).

In his later years after retirement, Glacken came less frequently to his office on the campus, and after the death of his wife Mildred became almost a recluse. The sequel to *Traces on the Rhodian Shore* remains a mass of papers, unpublished. He died in Sacramento on 20 August 1989.

2. Scientific Ideas and Geographical Thought

Glacken came late in life into academic geography, at an age when those who had entered postgraduate study directly following the BA could have reached the rank of full professor. Balancing the advantages of having a wide variety of experience against an early start and long career, he concluded he would have had a more limited and restricted outlook if he had followed the more usual course. It is fair to say that the variety and concreteness of his life experiences after graduating from the University of California as a social and intellectual historian brought him to the field of geography, and shaped and brought richness to his unique contribution to the man–nature theme in geographical thought.

Glacken's interest in the relation of man to his physical environment and in conservation began with his experiences with the migrants to the San Joaquin Valley who had lost their farms in the Dust Bowl. On his round-the-world trip he was conscious of erosion and abused land, as well as of the people and the varied landscapes. His association with and admiration for the work of William Vogt, and the 1949 UN conference on conservation of natural resources he attended sharpened his conception of the impact of growing populations on resources and on environmental degradation and prompted his entrance into postgraduate study in geography.

His interest in conservation and the relation of peoples to their environment developed along two lines of thought. He continued his focus on the history of ideas stemming from Teggart's course. He investigated writings across the ages for ideas about the relationship of man and nature. Ultimately, he identified and recognized the importance and antiquity of three leading ideas which he traced and described in *Traces on the Rhodian Shore*. He was also interested in how landscapes were created over time, in what people of different cultures did with their land according to their ideas and customs. The cultural geography course he developed with Paul Wheatley was organized round themes such as religion, gardens, land tenure, and technology and the impact of these immaterial aspects of culture on the land.

On his office wall Glacken had a favourite cartoon that epitomized his attitude toward landscapes. The picture showed the back gardens of two identical terrace houses. One is tidy, with the grass mown, borders trim, shrubs pruned, and flowers planted in neat beds. The other is messy and wild, with the grass long and the shrubs overgrown. In the first the couple are hard at work trimming, pruning and watering; the couple in the other garden are relaxing in deck chairs – two ideas of how to spend a weekend and of what is beautiful, with two different landscapes as a result.

Glacken's travel in Asia and Europe, visiting the sites of antiquity, and his residence and work in Korea and Okinawa furnished many examples of landscapes created by peoples of different cultures with long histories. He wrote to Carl Sauer from Oslo in January 1958:

> I became convinced . . . that in Europe, anything one chooses to study . . . must be essentially an historical study; that studies of contemporary life which relegate the history to pleasant prefaces are wrong if for no other reason than that they regard the present as something frozen enough to permit wide-ranging generalizations. The historical point of view also prevents the silly aping of science in attempts to find general laws. (Bancroft, Sauer Papers)

Glacken was then in Norway on a Fulbright Fellowship to study forest history and problems of conservation. He became convinced of the close relationship between the history of forests and the history of civilization. He was, however, less interested in the history of changes that man had made on the Earth than in the interpretations made of these changes. This was the theme of 'Changing Ideas of the Habitable World', the paper he presented at the ground-breaking, international symposium 'Man's Role in Changing the Face of the Earth' in 1955. For the Tenth Pacific Science Congress in 1961, 'Man's Place in the Island Ecosystem', he wrote 'This Growing Second World Within the World of Nature', outlining the historical background of modern ecological ideas. He was meanwhile pursuing his research, organizing his thoughts, and writing *Traces on the Rhodian Shore*.

This major work had its genesis in his entry into academic geography at Johns Hopkins and in his doctoral dissertation, 'The Idea of the Habitable World'. He knew when he entered Johns Hopkins that he wanted to write a history of ideas on a subject involving culture and environment. The dissertation was a history of several interlocking ideas from about the middle of the eighteenth century to the postwar years. He discussed the relation of population growth to resources, analysing Malthus's writings; the influence of the geographical environment on human culture, the idea that the physical environment had moulded human society; the idea of man as a geographical or geological agent, expounded in the writings of George Perkins Marsh and Nathaniel Shaler, leading up to the current pleas for conservation of natural resources; concepts of the soil; and living nature as a series of complex interrelationships.

Glacken developed and started teaching his own special course his first year at Berkeley. Derived from the dissertation, it was entitled 'The Relations Between Nature and Culture'. The course catalogue describes it as:

> A critical survey, from antiquity to the present, of leading theories of the effects of the physical environment on culture; the influence of these theories on such fields as geography, history, and anthropology; contemporary views of the nature of the physical environment in its relation to population and economic potentials. (Bancroft, University of California General Catalog, 1952–53)

He taught this course almost every year, and was called back twice after his retirement to teach it. In 1961 the course description was slightly altered, reflecting the gradual changes Glacken introduced into it as he delved into literature pursuing the ideas he had developed in his dissertation, where his analysis had been limited to the period from the mid-eighteenth century to the present. He came to realize the complexities of these ideas and that he must look for their origins in earlier ages and trace them through the writings of antiquity, the Middle Ages, and the Renaissance.

Carl Sauer had called the 'Man's Role in Changing the Face of the Earth' symposium a George Perkins Marsh Festival, as so many papers seemed inspired by his *Man and Nature*. Glacken was again impressed by Marsh's accounts of ancient environmental changes brought about by human action. Buffon's *Histoire naturelle*, which Glacken studied intensively, also had descriptions of environments altered by culture.

When he realized that man's abuse of the land has a long history, Glacken was inspired to look for recognition of man as a geographical agent in writings of the past, and discovered that that idea also has a long history. It became one of the three interlocking themes of *Traces on the Rhodian Shore*.

The second theme is environmental determinism, which as an idea also has a long history, predating Friedrich Ratzel, Ellen Semple and Ellsworth Huntingson. (Sauer had been one of the first American geographers to shake off the shackles of that thesis.)

Glacken confessed to David Hooson (*Geographers on Film*, July 1980) that he was late in recognizing the antiquity and importance of the third leading idea he discusses in his great book: the religious, teleological idea of the Earth as designed for man, a perfectly harmonious Earth and cosmos. He became convinced that this concept is the historical basis for what later became scientific and secular ecological thought.

The title of *Traces on the Rhodian Shore* was inspired by Vitruvius's story of the philosopher shipwrecked on the shore of Rhodes, who, seeing geometric figures on the sand, tells his companions, 'Let us be of good hope, for indeed I see the traces of men.' The geometric figures represent ideas, and the main, central tenet of Glacken's thought is that men's ideas become impressed on the land and show in the landscape.

In 1965, as Glacken was nearly finished writing *Traces*, he participated in J. Fraser Darling's 'Future Environments of North America', a major conference of scientists and conservationists. His paper, 'Reflections on the Man–Nature Theme as a Subject of Study', emphasized the importance of the role of ideas and attitudes about nature in forming public opinion, laws, political actions and scientific research.

Glacken intended to continue his history of the three themes as they occurred in the writings of the nineteenth and early twentieth centuries, and to go into another broad area that he 'would neglect only at my peril . . . the development of subjective, emotional, and aesthetic attitudes towards the natural world' (ibid.). This pursuit led him into reading the Romantic poets and other English literature as well as German, French, and American literature. He told Hooson that this section of the book was nearly finished but that there was an enormous body of material dealing with the other three ideas that he was not sure what to do with. This vast undertaking, to examine the thinking of the nineteenth century on the subject of man and nature, was never completed.

3. Influence and Spread of Ideas

Glacken made no apology for dealing almost exclusively with the ideas of Western civilization in his writings. He had studied ideas in non-Western cultures and found many parallels with Western thought, but he never considered doing scholarly work with non-Western sources because when dealing with ideas one must know how they are expressed in the original language (Buttimer, p. 30). He encouraged postgraduate students from non-Western countries to undertake such studies of ideas, and several have done so.

Glacken thought it important to introduce postgraduate students to the history of thought. For one thing, he told Allan Pred, it teaches a person humility to realize that 'new' ideas may be thousands of years old. Also, it gives the students confidence to know what they are dealing with, that there is a long history of their field, that leading ideas of social and geographical thought and theory date from the ancient world. However, he realized that intellectual history is 'not everyone's cup of tea' (Pred interview).

Glacken worked with many doctoral students at Berkeley, as either chairman or a member of their dissertation committee. Some did library research, delving into Swedish or Latvian attitudes toward nature, interpretations of California landscape, or the history of geographical thought. Others worked in the field, in New Guinea, the Philippines, or Taiwan, for instance, bringing the Glacken point of view to their interpretation of the landscapes. Still others worked on themes Glacken touched upon in his cultural geography course, such as tourism, gardens, geomancy or Asian cities. Like the teachers he admired from his school days, he was an encourager, making students sensitive to new ideas and encouraging them to develop in their own way. Because the cultural geography sequence was required for all postgraduate students, it was hard for them not to be influenced by his views, whether working directly with him or not, and when in the field not to be aware of the mark of different cultures on the landscape.

The course was not designed only for students intending to become professional geographers, but also for undergraduate students in general, for 'the little girl named Barbara', as Glacken used to say, with fondness and no disrespect. These students, as they travel throughout the world, may well have a deeper appreciation as they recognize and think about the cultures and ideas that form the landscapes they enjoy. The impact of a good teacher is like ripples in a pond, and Glacken was a good teacher.

Glacken's approach to the history of geographical thought is more subtle and difficult to comprehend than the usual facile practice of introducing the names of geographers and describing the major works, which are rarely read in the original. His work serves as a model, an illustration of a different and more profound and enlightening way to study the history of geographical ideas. *Traces on the Rhodian Shore* has been in steady, constant demand since its publication in 1967, and now, in 1991, is in its third paperback printing. It has become a classic, a basic text.

Bibliography and Sources

PUBLISHED WORKS BY CLARENCE J. GLACKEN

1939 'Medicine follows the crops', *Survey Midmonthly*, Vol. 75 (1939), 71–2. Joint author. A discussion of the medical programme of the Farm Security Administration for migrant labourers.

1940 'Tents, trailers, and culture', *Labor Information Bulletin*. (US Department of Labor, Bureau of Labor Statistics), Vol. 7, no. 8 (1940), 1–3. An informal discussion of daily life in the migratory labour camps.

1948 *Review*: *A Population History of the World*, *J. Econ. Hist.*, Vol. 11, pp. 74–5 (Marcal R. Reinhard, *Histoire de la population mondiale de 1700 à 1948*).

1951 'Control of climate in the home', *Geogr. Rev.*, Vol. 41, pp. 164–5.

1952 'The idea of the habitable world', in *Abstracts of Completed Doctoral Dissertations for the Academic Year 1950–1951*. Abstract Series No. 1, US Department of State, Washington, DC, pp. 32–5.

1953 *Review*: Heinrich Schmitthenner, *Studien über Carl Ritter*, *Geogr. Rev.*, Vol. 43, pp. 288–9.

1954 'The idea of progress and the habitable world', *Ann. Assoc. Am. Geogr.*, Vol. 44, pp. 209–10.

1955 *The Great Loochoo: A Study of Okinawan Village Life*. University of California Press, Berkeley and Los Angeles, xvi + 324 pp.

1956 *Review*: *Splendour of Earth: An Anthology of Travel*. Compiled by Margaret S. Anderson. *Geogr. Rev.*, Vol. 46, pp. 135–6.
Review: Jeannette Mirsky, *Elisha Kent Kane and the Seafaring Frontier*. *Geogr. Rev.*, Vol. 46, pp. 137–8.
'The origins of the conservation philosophy', *J. Soil and Water Cons.*, Vol. 11, pp. 63–6.
'Man and the Earth', *Landscape*, Vol. 5, no. 3, pp. 27–9.
'Changing ideas of the habitable world', in *Man's Role in Changing the Face of the Earth*, William L. Thomas, Jr. (ed.), University of Chicago Press, Chicago, pp. 70–92.
'Culture and the idea of nature', *YB Assoc. Pac. Coast Geogr.*, Vol. 18, pp. 23–7.

1957 *Review*: *The Reason of State*, translated by P.J. and D.P. Waley, and *The Greatness of Cities*, translated by Robert Peterson, Giovanni Botero. Introduction by D.P. Waley, Yale University Press, New Haven, CT, 1956 (Yale University Press Series, Rare Masterpieces of Philosophy and Science). *Landscape*, Vol. 6, no. 3, pp. 36–7.

1957 *Review*: *Places. A Volume of Travel in Space and Time. Places Which have Delighted, Intrigued, and Intimidated Man*. Edited by Geoffrey Grigson and Charles Harvard Gibbs-Smith. *Geogr. Rev.*, Vol. 47, pp. 285–6.

1959 *Review*: David Lowenthal, *George Perkins Marsh, Versatile Vermonter*. *Geogr. Rev.*, Vol. 49, pp. 437–8.

1960 *Review*: Arthur Koestler, *The Sleepwalkers. A History of Man's Changing Vision of the Universe*. *Landscape*, Vol. 9, no. 3, p. 35.
The Great Loochoo, reissued as a paperback by Chas. E. Tuttle Co., Boston, for sale in Asia.
'Count Buffon on Cultural Changes of the Physical Environment', *Ann. Assoc. Am. Geogr.*, Vol. 50, pp. 1–21.

1961 *Review essay*: *The City in History*, by Lewis Mumford. *Landscape*, Vol. 11, no. 1, pp. 33–5.

1962 *Review*: *Friedrich Ratzel, a Biographical Memoir and Bibliography*, by Harriet Wanklyn. *Geogr. Rev.*, Vol. 52, pp. 467–8.

1963 'This growing second world within the world of nature', in F.R. Fosberg (ed.), *Man's Place in the Island Ecosystem* (Honolulu, HI: Bishop Museum Press), 10th Pacific Science Congress, Honolulu, 1961, pp. 75–95.

1964 *Review*: *The Domesday Geography of Southeast England; the Domesday Geography of Northern England*, edited by H.C. Darby *et al. Geogr. Rev.*, Vol. 54, no. 2, pp. 285–7.

1965 Introduction to *The Geographical Lore of the Time of the Crusades*, by John K. Wright, pp. vii–xiii, Dover Publications, New York.

1966 'Reflections on the man–nature theme as a subject of study' in John Fraser Darling (ed.), *Future Environments of North America*.

1967 *Traces on the Rhodian Shore*. University of California Press, Berkeley and Los Angeles, xxviii + 763 pp. (reprinted 1973).

1969 'On Chateaubriand's journey from Paris to Jerusalem, 1806–07', in *The Terraqueous Globe*, W.A. Clark Memorial Library, UCLA, pp. 45–75 (Papers read at the Clark Library Seminar, 27 April 1968).

1970 'Man against nature: An outmoded concept', in *The Environmental Crisis*, H.W. Helrich, Jr. (ed.), Yale University Press, New Haven, pp. 127–142. (Reprinted by Warner Modula Publications, 1972.)

1970 'Man and nature in recent western thought', in *This Little Planet*, Michael Hamilton (ed.), Scribner's, New York, pp. 163–201.

1973 'Environment and culture', in *Dictionary of the History of Ideas*, Scribner's, Vol. II, pp. 127–34.

1973 *Traces on the Rhodian Shore*, reprinted in the California Library Reprint Series, University of California Press, Berkeley and Los Angeles.

1974 *The Great Loochoo*, reprinted by Greenwood Press, Westport, CT.

1976 *Traces on the Rhodian Shore*, University of California Press, paperback edition.

In preparation: *Culture and Environment in Western Civilization during the Nineteenth Century*, a sequel to *Traces on the Rhodian Shore*.

SOURCES

Published Sources

Glacken, Clarence J., 'A late arrival in academia', in Anne Buttimer (ed.), *The Practice of Geography*, London, New York, Longman, 1983.

Gardner, David, *The California Oath Controversy*, Berkeley and Los Angeles, University of California Press, 1967.

Rorabaugh, W.J. *Berkeley at War: The 1960s*, New York, Oxford, Oxford University Press, 1989.

Dow, Maynard W. (ed.), *Geographers on Film: Clarence Glacken*, Interviewed by David Hooson, July, 1980.

Unpublished Sources

University of California, Department of Geography

Biographical Sketch, Clarence J. Glacken, 15 February 1952.

Curriculum Vitae, Clarence J. Glacken (n.d.).

Biography for Academic Personnel, Clarence James Glacken, 10 October 1978.

Videotaped interview with Clarence Glacken by Allan Pred.

Bancroft Archive

President's File,
Chancellor's File
Social Science Council
General Catalog of Courses, 1949–1976
Carl Sauer Papers
Margaret Hodgen Papers

Anne Macpherson is an independent scholar residing at 800 Park Way, El Cerrito, California 94530.

Chronology

1909	Born, 30 March
1926–28	Sacramento Junior College
1928–31	University of California, Berkeley
1930	BA, Department of Social Institutions
1931	MA, Department of Social Institutions
1931	Research Assistant, Department of Social Institutions
1934–36	California State Emergency Relief Administration California Resettlement Administration
1937	Eleven-month trip round the world
1938–39	Works Progress Administration, Oakland
1939–41	California State Relief Administration, Tulare
1941–46	Service in US Army
1943–44	Civil Affairs Training, Japanese Language School, University of Chicago
1945	Taught, Civil Affairs Training School, Chicago Campus, Northwestern University
1945–46	South Korea, Deputy Director of Bureau of Public Health and Welfare
1946	Taught, Carlisle, Pa., in Army School for Government of Occupied Areas
1946–49	Washington, DC, Veterans Administration, Office of Vocational Rehabilitation and Education
1948–49	American University, Lecturer in history
1949–51	Graduate Study, Isaiah Bowman School of Geography, Johns Hopkins University; Doctor of Philosophy, June 1951
1951–52	Field Associate in Cultural Geography, Pacific Science Board, National Research Council; Study of village life on Okinawa

1952	Joins faculty, Department of Geography, University of California, Berkeley; lecturer
1953	Assistant professor
1955	Wenner Gren Foundation International Symposium, 'Man's Role in Changing the Face of the Earth'; Paper: 'Changing Ideas of the Habitable World' *The Great Loochoo* published
1957–58	Fulbright Scholar in Norway to study problems of conservation
1958	Associate professor
1961	Tenth Pacific Science Congress Symposium, Honolulu: 'Man's Place in the Island Ecosystem'; paper: 'This Growing World within the World of Nature'
1964	Full professor
1965	Fellow, John Simon Guggenheim Memorial Foundation 'Future Environments of North America' Conference; paper: 'Reflections on the Man–Nature Theme as a Subject of Study'
1966	Chairman of Department of Geography
1967	*Traces on the Rhodian Shore* published
1967–68	Travel in Europe to study the history of the nature protection movement
1968	Association of American Geographers Citation for Meritorious Contributions to the Field of Geography
1970	Nervous and physical breakdown
1974	Heart attack
1974–75	Sabbatical
1976	Emeritus professor
1976	Berkeley Citation for distinguished achievement and notable contributions to the university
1989	Died, 20 August

François Emile Matthes

1874–1948

Fritiof Fryxell and
William W. Speth[1]

The place which François Emile Matthes holds in American earth science is secure and, in certain respects, unique. He stands alone in having attained eminence first as topographer and later as geologist and physical geographer, and is one of the relatively few whose writings rank as both science and literature. The superb artistry and thorough understanding of land forms displayed in his delineation of terrain by the contour line led to the production of maps of such difficult and spectacular areas as the Bighorn Mountains of Wyoming, the Grand Canyon of the Colorado River, the Yosemite Valley, and Mount Rainier National Park. His ever-increasing interest in the origin of upland glacial features resulted in numerous published studies of glaciation and the physical processes of high altitudes, and led directly into the fields of climatology and meteorology. His colleagues held him in the highest esteem for the originality and importance of his contributions to science, and he was honoured by, and was a member of, many scientific societies.

1. Education, Life and Work

François Emile Matthes and Gerard Hendrik Matthes, twin brothers, were born in Amsterdam, the Netherlands, on 16 March 1874. Their parents both belonged to old, distinguished Dutch families. Willem Ernst Matthes, the father, was a well-to-do merchant, a member of the firm Matthes and Bormeester, dealers in colonial

[1]This biobibliography is based very considerably upon 'Memorial to François Emile Matthes (1874–1948)', *Proc. Geol. Soc. Am., Ann. Rep. for 1955* (1956), 153–68, by Fritiof Fryxell, for whom Matthes had been an inspiring personal friend. Before his death, the senior author, who was Matthes' literary executor, had agreed to a recasting of the Matthes memorial for an audience of geographers (Fryxell to Speth, personal letter, 10 June 1980).

products such as rubber, indigo, and hemp, the firm that had the contract to furnish India rubber for the first transatlantic cable, laid in 1866. He had travelled widely throughout Europe and was a man of broad culture and many interests. The mother, Johanna Suzanna Matthes, née van der Does de Bije, was a refined woman, with a keen mind and a flair for history. She was brought up in the old university town of Leiden, and mingled in court circles at the Hague.

When about eight years of age, the twins began to be troubled with malaria. The disease was common in Amsterdam, but since its cause was not understood no measures were taken to check the breeding of mosquitoes in the canals. The family doctor recommended for the boys a sojourn of several years in the Alps, which were known to be malaria-free. Accordingly, when the twins were nearly ten, they were taken by their mother to Switzerland, there to lay the foundations of health and to gain a cosmopolitan education - an event which was a turning-point in their lives.

At Courgevaux, a few miles from Lac Morat, the mother found a school suitable for foreigners, and they resided for more than a year in a wing of a nearby castle. There the twins learned to speak French fluently, and picked up a smattering of German, and some English. When they were eleven they lived first at Montreux in a first-class hotel overlooking Lake Geneva, and later in a smaller hotel in the vicinity. The boys' father, whose hobby was mountaineering, spent a week or two with them each summer. He taught them to read military maps (General Dufour's series), which were beautifully finished with hachures. Becoming skilled in the use and interpretation of topographical maps as boys was to prove of tremendous value to the twins in later years. Themselves patrons and collectors of art, the parents saw to it that from the age of nine the twins received instruction in drawing with pencil, charcoal, and ink. François revealed a native talent for art which amazed everyone. Beginning also at the age of nine, they were taught to draw on tracing linen, using triangles, T-square, and drawing pens.

In 1887 the twins, now thirteen, procured German grammars and began serious study of the German language. That autumn their mother took them to Frankfurt-am-Main, Germany. Their studies now continued at a technical high school, affording, besides German, excellent instruction in both French and English. From the first, travel was an important ingredient in the education of the Matthes twins. Now, a refreshing change from school life in Germany was provided by the summer months, when they visited their home in Amsterdam; travelled through France, Germany, and Switzerland; or stayed at the summer residence of the family at Baarn (near the beaches and dunes of the Netherlands), or visited the country estate near Brussels, Belgium, of a close friend of the Matthes family.

The studies at Frankfurt were intended as preparation for an engineering course at a German university; but after the twins had spent four years in Germany, their plans underwent another momentous change. Instead, in September 1891 they emigrated to *Amerika*. After reaching the United States, they sought admission to the Massachusetts Institute of Technology, but the semester was already underway. However, the president, General Francis A. Walker, became interested in the boys and permitted them to take the entrance examinations. These showed them to be far ahead of requirements in modern languages and mathematics, and deficient only in American History. They were admitted at once to the civil engineering course, as freshmen, but were excused from taking French and (being foreigners) from military drill. With the writing of English, their fourth language, they had little difficulty; and their rapid progress in the mastery of spoken English is evidenced by the fact that they were soon giving talks to the Civil Engineering Society of the Institute and other groups, on various technical subjects. Their early training in drawing and drafting now stood them in good stead; François soon became the outstanding draughtsman in his class. On the advice of a professor who had formerly been connected with the US Coast and Geodetic Survey, the boys elected the geodetic course, which would prepare them for careers with that Survey. Their third summer in the United States, 1894, was spent at the MIT field school in the Adirondack Mountains, where they learned mapping with plane-table and alidade, and, in the Ausable River, measured stream flow with different kinds of current meters. In 1895 the twins both graduated with honours, receiving the degree of Bachelor of Science, and the next year they became US citizens.

Following graduation, the twins, Gerard and François, each went his way to climb to the top of his respective profession. Gerard distinguished himself in the field of hydrographical engineering; François for many years expressed his talent for drawing, as well as for precise, analytical, scientific work, in the making of topographical maps. Eventually, however, he became a geomorphologist and a glacial geologist. As the latter specialty came to include the study of existing glaciers, it, in turn, led him far into the cognate field of atmospheric science. Thus, François Matthes reached early maturity in an age in American academic life when would-be students of earth science were obliged to study geology or physiography; it is impossible to disentangle the development of geomorphology or, for that matter, cartography, which was essentially topographical up until the 1930s, from that of physical geography in the United States in the decades spanning the turn of this century.

In the same year as his graduation, Matthes commenced his professional career as draughtsman and instrument man in the city engineer's office at Rutland, Vermont, and was assigned to the making of detailed topographical surveys for the city. Exactly one year later he joined the US Geological Survey, the organization with which he was to be connected for fifty-one years. Appointed a field assistant in the Survey in 1896, he advanced to assistant topographer in 1898, a grade which qualified him for the larger opportunities that were to materialize. Thenceforth he was promoted to topographer in 1899, topographical inspector in 1907, associate geologist in 1913, and geologist in 1916. From 1930 until his retirement in 1947 he was senior geologist.

His first important assignment as a topographer was as chief of the party which was charged with surveying the Cloud Peak quadrangle, the most rugged portion of the Bighorn Mountains of Wyoming. The lower, western third of the quadrangle had been surveyed in 1897; it remained for Matthes to survey the remainder of the area in 1898 and 1899. Of Matthes' work as a topographer, S.S. Visher noted that 'he is credited with having contributed notably to the effectiveness of mapping rugged mountain areas', and that his method, which involved multiple sights through the telescopic alidade and the sketching of provisional form lines, was to become standard procedure (Visher, 1948, p. 301). (Prior to about 1935, all the topographical maps produced by the Geological Survey were made with the use of the alidade and plane-table.) Matthes' success as a topographer stemmed not only from mastery of technical principles and methods, and from his skill in drawing, but also from his keen analysis of land forms and his consuming desire to understand them. The Bighorn Mountains provided him with his first opportunity to delineate, with contour lines, the varied forms of mountain landscapes, especially those resulting from alpine glaciation; and as he brought out the topographical features on his maps he pondered over their origin. Thus, following his fieldwork, in 1899, he rendered a service beyond the call of duty as topographer when, in the evenings, after long days of drafting at the Washington, DC, office, he wrote *Glacial Sculpture of the Bighorn Mountains, Wyoming* (1900). This was his first scientific publication, and it was still a standard reference more than fifty years later.

From his initial assignment in the Rocky Mountains, Matthes emerged a seasoned topographer of recognized skill, and a first-rate horseman and back-packer, with experience which was to prove all but indispensable to him in his future work. In the spring of 1900 he made a hydrographical reconnaissance of the Blackfoot Indian Reservation in Montana. There followed a succession of assignments which included two areas in Montana (summers of 1900 and 1901), two areas in Arizona (winters of 1900–01 and 1902–03), and several other places which, from the standpoint of the topographer, were as difficult and spectacular, and therefore as challenging, as any to be found in the entire Far West. That he became a master of reconnaissance mapping – first topographical and later geological, much of it carried out in exceedingly difficult terrain without benefit of roads, or, for that matter, even trails, at a time when modern aids to mapping such as aerial photography were unheard of – may be attributed not only to hardihood and perseverance but also to his consummate skill in planning and executing long pack-animal trips, and to his great resourcefulness in solving the problems posed by exploration in wilderness country.

In February 1902 came orders to begin the mapping of the upper half of the Grand Canyon of the Colorado River, in Arizona, the scenically more remarkable section of the canyon which was destined later (1919) to become Grand Canyon National Park. Aside from the difficulties of getting about in, and mapping, some 500 square miles of almost incredibly rugged country and of maintaining the life of men and animals in a forbidding, practically uninhabited region, largely devoid of usable water supplies, the assignment was a staggering undertaking. Suffice it to say that the mapping was done wholly by means of the plane-table, the Grand Canyon providing, in Matthes' words, 'an extreme test of the efficiency of that instrument'. It was the original intention to publish the maps in fifteen-minute quadrangles, on the standard scale of 1:62 500 and with one-hundred-foot contour intervals; but on the recommendation of Matthes, in order to bring out the distinctive topography of the area, the Grand Canyon quadrangles were made an exception to usual practice and published on the scale of 1:48 000, with fifty-foot contour intervals.

The mapping of the Grand Canyon continued in 1903, and only in January 1904, after nearly two years of practically continuous fieldwork, did Matthes leave by train for Washington, DC. There remained the task of inking the maps and supervising the engraving, eye-straining work which permanently impaired Matthes' sight. The end-products of this labour, however, the Bright Angel and Vishnu quadrangles, stand as examples of topographical excellence which have never been surpassed. Indeed, the delineation of topography by isohypses in the early years of the Geological Survey was, in the words of John Leighly, 'seen at its best in François Matthes's Bright Angel quadrangle of the Grand Canyon, and especially in his map of Yosemite Valley. . . . Such maps as Matthes's are the product of free, gratuitous art, as so much of the writing done under the Survey represents free and gratuitous science, superimposed on the official and formal inventory of the public lands. They are among the finest expressions of the esthetic motive in our [American] tradition of geography' (Leighly, 1971, p. 6).

In the winter of 1904–05, Matthes took a postgraduate course in geology at Harvard University, majoring in geomorphology under Professor William Morris Davis. While at Harvard he also lectured on 'topographical field methods for advanced students in geology', under an Austin Teaching Fellowship. During this time the paths of Isaiah Bowman (1878–1950) and François Matthes crossed; they were partners in an advanced geology field course, and, subsequently, Bowman worked with Matthes during one field season for the Geological Survey (Martin, 1980, 12–14). Matthes had almost fulfilled requirements for the MS degree when there came the offer from the Survey to make a large-scale (1:24 000) map of the Yosemite Valley. (Matthes had dreamed of Yosemite ever since coming to *Amerika*, and, in June 1903, he interrupted his work at the Grand Canyon to travel by train and stagecoach to Yosemite National Park. He beheld it at the height of its springtime beauty, and he was never to forget it.) This was an opportunity he could not forgo, but he enquired if he might not be allowed the few weeks of university work necessary for completion of his course. The reply was negative. He hesitated no longer, for the chance to map the Yosemite Valley was worth far more than an additional degree, and on 14 June 1905, he stood once more on the floor of 'the incomparable valley'. Two seasons were necessary to complete the field-work on the Yosemite map. In October 1906, Matthes began inking the map at the Sacramento office of the Survey. The engravers in Washington, DC, completed the task, and in July 1907 the 'Yosemite Special' was an accomplished fact. In 1923 Matthes wrote an interpretive text for the reverse side of the map - one of a number of such essays which, at various times, he wrote for the maps of National Parks and other areas.

Matthes spent the spring of 1906 in Berkeley, California, assisting Grove Karl Gilbert (1843–1918) in a study of the transportation of sediment by rivers, aiding both with the laboratory experiments and by reviewing the extensive Dutch literature relating to silt deposition in the rivers of the Netherlands. On 18 April occurred the great San Francisco earthquake, which caused more than an unforeseen interruption of their scheduled work. Gilbert was appointed to the California State Earthquake Investigation Commission, and Matthes, assigned to the field almost immediately on behalf of the Commission, traced and mapped the San Andreas Rift through the northern half of the state. He also made a special, detailed map of the rift in the vicinity of Fort Ross, in Sonoma County. His maps were published in the atlas prepared by the Commission, and his field observations, relating particularly to the fault as a geomorphical feature, were embodied in the Commission's two-volume report.

From 1907 to 1913, Matthes served as an Inspector of Maps for the Geological Survey, visiting widely scattered field parties during the summer months. In winter, in Washington, DC, he supervised the work of topographers, as the latter completed the drafting and inking of their field manuscripts. In January 1908 Matthes and four associates founded the Quadrangle Club, for purposes of instruction, and for the exchange of experience and views. Matthes was a leading spirit in planning these sessions, which continued until 1917, and also in giving papers at them.

In 1910 came Matthes' last major field assignment for the Topographic Branch, when he was put in charge of the mapping of yet another National Park, Mount Rainier, in the State of Washington. Once more he found himself working among glaciers, in this case the largest number to be found on any peak in the United States. The rock cleavers between the glaciers offered a challenge to map maker and mountain climber alike. Inevitably, his interest in glaciers and their work received fresh stimulus, and he again wrote several geological papers as the by-products of a topographical assignment. Because of poor visibility, resulting from bad weather and forest fires, the field seasons of 1910 and 1911 permitted Matthes to complete little more than a fourth of the map (the southwestern part). The remaining part was completed in 1913 by others.

On 7 June 1911, François Matthes married Edith Lovell Coyle, a native of Washington, DC, and a descendant of some of the founders and prominent citizens of the Virginia and Maryland colonies. Among the interests which she and François had in common were love of the out-of-doors and appreciation of the beautiful in nature; she was to accompany her husband in future field seasons, whenever possible, and to aid him in his work.

Matthes' concern with geomorphology had grown, year by year, and he had written and lectured extensively in this field. On 1 July 1913, he was transferred from the Topographic Branch of the Geological Survey to the Geologic Branch, where he could devote himself wholly to geomorphology. Though Matthes was now nearing forty, he welcomed the change as one which would enable him to give undivided attention to geological problems, which by this time had become his primary interest. The many years he had devoted to topographical mapping gave him an unusual and, from many standpoints, an enviable background for the second phase of his career.

Matthes' first geological assignment was to study the origin of the Yosemite Valley and, what was without precedent, to report his findings in language understandable not only by scientists but by any intelligent and interested readers. Not until sixteen years later, in the autumn of 1930, did his completed report, *The Geologic History of the Yosemite Valley*, Professional Paper 160, come off the press. In reviewing this most widely acclaimed of Matthes' publications, Kirk Bryan wrote, 'Occasionally in the history of science there appears a work so excellent, so comprehensive, that it becomes immediately a classic. Such a newborn classic is the long-awaited "professional paper" by Matthes on the Yosemite Valley' (Bryan, 1932, 84). In Professional Paper 160 Matthes goes far beyond the limitations set by his title, for he interprets the Yosemite Valley in its relation to the geological history of the entire Sierra Nevada, giving special emphasis to problems of structure, geomorphology, and glacial geology.

World War I brought Matthes special assignments, such as describing the geological environments of Camp McClellan, Alabama, and Camp Gordon, Georgia. In 1919 his attention returned to the Sierra Nevada for many years of further work. Professional Paper 160, though a large volume, did not contain all the data which Matthes had gathered in the Sierra Nevada, for from the Merced and Tuolomne basins he had extended his investigations southward into the upper San Joaquin basin and even as far as the Kings Canyon and Sequoia National Park. Through these reconnaissances he had become engrossed in the larger problems of Sierran geology, and he was eager to pursue these further.

But there followed the 'Mississippi Valley interlude', from 1928 to 1934, during which he was assigned to geological problems in the Middle West. These put him far behind in his Sierran work – one reason why a number of his Sierran studies were never finished. In 1935 he returned to California, this time through a co-operative agreement with the National Park Service, for a reconnaissance of Sequoia National Park. The Sequoia data were urgently needed by the Park Service, and to meet this need Matthes, pending the opportunity to write a formal report, assembled three volumes which summarized the geology of the park by means of annotated photographs. He was able to complete only one other major paper on Sequoia National Park, namely, *The Geologic History of Mount Whitney*, published in 1937. In 1937 Matthes continued fieldwork in Yosemite National Park and late in the summer began to investigate the eastern front of the Sierra Nevada. The resulting study he came to regard as his foremost geological contribution, for in it he concluded that the great eastern escarpment of the range had been formed as a result of early Pleistocene faulting, and hence was of much more recent origin than had been supposed. In 1938 and 1939 he was able to round out his evidence, at least in part. In July 1939 he presented his findings at the meetings of the Cordilleran Section of the Geological Society of America, in Berkeley; that year was destined to see his last field excursion into the Sierra Nevada.

Although in his later years Matthes published many papers, his time and strength were increasingly consumed by the demands of various scientific organizations. As chairman of the Committee on Glaciers of the American Geophysical Union, he organized and supervised a programme for conducting co-operative studies and measurements of the existing glaciers in the US and abroad, a work which made him an international leader in deciphering the record of post-Pleistocene, prehistoric fluctuations of climate. Also, Matthes prepared detailed reports summarizing and analysing the data on glaciers, particularly with reference to their year-to-year changes, and these reports were published annually from 1932 to 1946 in the *Transactions* of the American Geophysical Union.

Matthes returned to Washington, DC, in the autumn of 1939 when the war in Europe had begun. Delegates to the triennial assembly of the International Association of Scientific Hydrology were convening in the nation's capital. When the French secretary departed abruptly for his native land, Matthes – the only available linguist in the American delegation – was drafted to take his place. So he was burdened, not only at these meetings but for years afterward, with the responsibility for the activities of this group, one of the largest international scientific organizations. He was also made secretary of the International Commission on Snow and Glaciers, another division of the Association; this position likewise called for an extraordinary amount of correspondence with colleagues throughout the world.

On top of all this, Matthes was invited to write a chapter on *Glaciers*, for the volume on *Hydrology*, Number IX of the *Physics of the Earth* series, of the National Research Council. Review of the worldwide literature on glaciers and the writing of the chapter, which itself grew to the size of a book, took practically all of 1941. This treatise, a masterly summary of the behaviour and characteristics of glaciers, Matthes himself considered his principal contribution to glaciology.

During World War II, Matthes was called on to make contributions to the war effort, including the work of the Military Geology Unit of the Geological Survey, by reason of his broad geographical knowledge and his command of foreign languages. In 1945 he undertook a critical re-examination of W.H. Hobbs' doctrine of glacial anticyclones, which he regarded as scientifically untenable and indeed dangerous in that it afforded a false basis for the use of Greenland as an airbase. In Part I of this study, published in 1946, he concluded that 'there is no evidence of a virtually permanent "glacial anticyclone" centered over the Greenland ice sheet. On the contrary . . . weather over the ice sheet is controlled by alternating cyclonic and anticyclonic movements. . . . Aviators need not expect so much as a fifty-fifty chance of meeting with good weather flights across central Greenland, except for a few weeks in mid-summer.' Part II of this study was completed, after Matthes' death, by Arthur D. Belmont.

Because of heavy demands made on the Geological Survey during and after the war, Matthes was continued on the active roll somewhat more than three years beyond the statutory age of retirement (seventy years), but on 30 June 1947, he retired. In September, François and Edith drove across the country to their retirement home in El Cerrito, California.

Matthes gave himself feverishly to the task of arranging his library and professional materials, and in February 1948, with long-deferred writing projects scarcely begun, he was handed a tremendous organizational task: to plan the programme of the Committee on Snow and Glaciers for the sessions of the International Scientific Congress, scheduled to meet in August, in Oslo. He felt he would be negligent of his duty if he did not accept this task which, he knew, would require all his time and energy for months to come, but he did so with a heavy heart. In the early morning hours of Sunday, 18 April, Matthes was stricken with a heart attack, and nine weeks later, on 21 June, he died in his El Cerrito home. Later, in accordance with the wish he had expressed, he was cremated and his ashes were scattered in his beloved Yosemite.

In the course of his life recognition came to Matthes from many sources. He was starred in *American Men of Science*, and he served as president of the Geological Society of Washington in 1932, vice-president of the Washington Academy of Sciences in 1933, and president of the Association of American Geographers in 1933. Upon learning of the last-named honour, Carl Sauer remarked in a letter to his friend Vernor C. Finch at the University of Wisconsin, 'I think the election of Matthes to the presidency is rather nice. He has done some mighty good work and is a real scholar' (C.O. Sauer to V.C Finch, 19 December 1932, Carl O. Sauer Papers, Bancroft Library, University of California, Berkeley, California). Matthes was one of the forty-eight original members of the Association and served as treasurer from 1912 to 1919 and as a member of the Committee on Physiographic Provinces (James and Martin, 1978, 37 ff.). His presidential address, 'Our Greatest Mountain Range, the Sierra Nevada of California', was not published, but on the basis of temporal juxtaposition it is reasonable to infer that much of its substance is reproduced in the article 'Geography and Geology of the Sierra Nevada', which appeared in the *Guidebook* published in conjunction with the Sixteenth International Geological Congress in 1933. Matthes was also honorary vice-president of the Sierra Club, which, in 1949, took measures to give the names 'Matthes Crest' and 'Matthes Lake' to two suitable geographical features in Yosemite National Park. The long list of his affiliations with learned societies, mountaineering clubs, and other organizations, both domestic and foreign, can be found in *The National Cyclopaedia of American Biography* (1950), and in *Who's Who in America* (1948–49), and need not be repeated here. Four honours may be mentioned as representative of the varied facets of his interests and work.

He was created a Chevalier, Order of Leopold II, by King Albert of Belgium, in 1920, in appreciation of the account which he gave to the King and his party of the origin of Yosemite Valley, when they visited Yosemite in the autumn of 1919. Matthes' avocation was in Boy Scout work, and he was awarded the Order of the Silver Beaver in 1931, by the Council of the Boy Scouts of America, for 'distinguished service to boyhood'. This was the honour he cherished most dearly. At the commencement exercises in the great stadium of the University of California in 1947, he was given the honorary degree of LL.D. In conferring the degree upon Matthes, President Robert Gordon Sproul said, in part, 'by your artistry in the delineation of land forms and your clear, scientific description, you have interpreted the beauty of the Western American Landscape to the mind as well as the eyes of all who love mountains.' Finally, on 28 April 1948, he was awarded the gold medal of the Department of the Interior for fifty-one years of distinguished service to the US government.

Both physically and intellectually, Matthes was impelled by a tremendous inner drive and tenacity of purpose; and his ideals with respect to workmanship were so high he was never willing to compromise short of the best results he could achieve. He was meticulous with respect to details, and conscientious and orderly to an extraordinary degree. Dignified and courteous, he seemed shy and aloof to some, but among kindred spirits he glowed with enthusiasm. He was a born teacher and might have accepted the teaching opportunities that came to him repeatedly had he not found the riddles of Sierran geology irresistibly fascinating. The aesthetic, furthermore, played a dominant part in his life; he was not only a scientist, but a poet, and an artist to the core. To a friend he confided, 'I simply cannot be happy unless my work is conducted in country of inspirational beauty.' During the 'Mississippi Valley interlude' he wrote to W.M. Davis, 'I am temperamentally not well fitted for work on the monotonous flats of the Mississippi valley. . . . I should much prefer returning to the mountains and carrying on there as long as I am physically fit, – the more so if, coupled with the mountain work, there should be opportunity for contact with the public.' Sharing his scientific findings with the public – as in the national parks – delighted him. He was always mindful of 'those who love the mountains, particularly those who come to see, and seeing, wonder and wish to understand'.

2. Scientific Ideas and Geographical Thought

Unlike his Harvard classmate Isaiah Bowman, François Matthes did not explore and elaborate in his work his teacher's principle of ontography. Matthes remained preponderantly the student of the delineation and interpretation of topographical forms, mainly in the mountainous West. In the first of his many publications on glaciation and its effect on topography, *Glacial Sculpture of the Bighorn Mountains, Wyoming*, Matthes distinguished from the land forms sculptured by glaciers other features produced by persistent snowfields, and he applied the now-familiar term 'nivation' to the previously unrecognized geomorphical processes which were involved. In a review of the monograph, R.D. Salisbury (1858–1922) wrote, 'This, so far as we are aware, is the first attempt to analyze the effects of inactive ice and névé on topography' (Salisbury, 1901, 465).

As chairman of the committee on glaciers of the American Geophysical Union, Matthes began assembling a collection of dated photographs and other records relating to the existing glaciers in the US, chiefly in the period from 1931 to 1948. This file has grown through the years and has become increasingly valuable for reference, revealing the changes in extent and volume which the glaciers have undergone; now held by the US Geological Survey, this collection already has considerable historical significance. His chapter on *Glaciers*, however, expressed some of the ideas which had long been maturing in his mind, such as the view, now widely accepted, that most of the glaciers in the western US are not the remnants of Pleistocene ice bodies, as previously assumed, but instead are 'modern' glaciers which became re-established in post-Pleistocene time, probably as recently as within the last 4000 years, in what he aptly termed the 'little ice age'.

3. Influence and Spread of Ideas

Matthes' scientific stature and influence at the time of his retirement can be gauged, in part, by the relatively high frequency with which his writings are cited in Richard Foster Flint's landmark volume *Glacial Geology and the Pleistocene Epoch* (New York, Wiley, 1947). Eight bibliographical citations, which span a period of forty-three years, and numerous textual references by Flint attest to the effect of Matthes' intellectual output on his profession. Furthermore, his topographical studies in Wyoming (the Cloud Peak quadrangle) and Montana (the Chief Mountain quadrangle) made clear to his readers the existence of an American 'Alps'. Conservation was to the fore, enthusiasm grew, and the superb Chief Mountain map, together with Matthes' photographic records and his publications, gave telling impetus to the movement which culminated in 1910 in the act of Congress establishing Glacier National Park.

During the time he spent in Washington, DC, after 1908, Matthes became a leading spirit in planning the many meetings which took place within the Topographic Branch of the Geological Survey, for purposes of instruction, and for exchange of experiences and views. His own discourses at these sessions brought new and stimulating ideas to aspiring topographers, which deepened their understanding and appreciation of the science of topographical mapping. These beliefs were the outgrowth of his own rich experience and of his acquaintance with mapping practices abroad and the current European literature. In his talks he reiterated his conviction that 'topography is something more than engineering'; it partakes of 'an interpretative and synthetic art'. He insisted that 'no topographer who would aspire to give maximum value to his work can hope to do so if he lacks an intelligent insight into the nature of land forms.'

Matthes' interest in topographical mapping remained keen, even after he was transferred in 1913, to the Geologic Branch of the Survey. In 1919 he presented a paper before the Washington Academy of Sciences on 'Relief shading on topographic maps', a subject very near his heart and one in which he had long sought to arouse interest. When, eventually, the Geological Survey began publishing shaded topographical maps as standard practice, it was a source of great satisfaction to him.

As an interpreter of the Western scene, Matthes was without equal among contemporary American earth scientists. Even when writing technical papers he kept in mind the interested laymen who shared his interests. Matthes' attempt to address the widest possible circle of readers, by reducing the use of scientific jargon, was so successful that it was emulated by many other writers; *The Geologic History of Yosemite Valley*, Professional Paper 160, remains the model in its field. It is not surprising, therefore, that the demand for this volume has exceeded, as of the late 1950s, that for any other Professional Paper. Copies of the first printing are now coveted by collectors of Californiana.

Although his concern with the general public netted him little in terms of professional standing, Matthes' eagerness to impart to others what he himself perceived in landscapes led him, very early in his career, to contribute geological essays to mountaineering and other journals. Some of his finest essays were written about the national parks and were printed on the reverse side of the topographical maps of these areas. This use assured them an exceptionally wide distribution, and it put them in the hands of those most likely to read them with profit and appreciation. Most of Matthes' essays, however, were written for the *Bulletin* of the Sierra Club, and were addressed primarily to his fellow members in that group. 'Nothing makes me happier,' wrote Matthes in one of his letters, 'than when I can share the results of my scientific studies with laymen.'

Bibliography and Sources

1. REFERENCES ON FRANÇOIS E. MATTHES

Salisbury, R.D., Review of *Glacial Sculpture of the Bighorn Mountains, Wyoming*, by F.E. Matthes, *J. Geol.*, Vol. 9 (1901), 465–6.

YB. Am. Alp. Club, 1923, 1932, and 1940. (Contain summaries of Matthes' topographical and geological explorations, and list his principal mountain ascents.)

Bryan, K., Review of *Geologic History of the Yosemite Valley*, by F.E. Matthes, *J. Geol.*, Vol. 40 (1932), 84–7.

Cattell, J. (ed.), 'Matthes, François E(mile)', in *American Men of Science* (1944), Lancaster, PA, Science Press, 7th edn, 1185–6.

——, 'Matthes, François Emile', in *Who's Who in America* (1948–49), Chicago, A.N. Marquis Co., Vol. 25, 1608.

McHenry, D.E., 'François E. Matthes honored – retires', *Yosemite Nature Notes*, Vol. 27 (1948), 54–7.

Visher, S.S., 'François Emile Matthes, 1874–1948', *Ann. Assoc. Am. Geogr.*, Vol. 38 (1948), 301–4.

Farquhar, F.P., 'François Emile Matthes, 1874–1948', *Am. Alp. J.*, Vol. 7 (1949), 201–3.

Ray, L.L., 'Obituary: François Emile Matthes', *Geogr. Rev.*, Vol. 39 (1949), 157–8.

——, 'Matthes, François Emile', in *The National Cyclopaedia of American Biography* (1950), New York, James T. White and Co., Vol. 36, pp. 39–40.

Fryxell, F., 'Memorial to François Emile Matthes (1874–1948)', *Proc. Geol. Soc. Am., Annual Report for 1955* (1956), 153–68.

Leighly, J., 'The "American tradition" in geography', *Minn. Geogr.*, Vol. 23 (1971), 2–10.

James, P.E. and Martin, G.J., *The Association of American Geographers, the First Seventy-five Years, 1904–1979* (1978), Washington, DC: Association of American Geographers, 279 pp.

Martin, G.J., *The Life and Thought of Isaiah Bowman* (1980), Hamden, CT, Shoe String Press, 272 pp.

2. SELECTIVE AND THEMATIC BIBLIOGRAPHY OF WORKS BY FRANÇOIS E. MATTHES

a. Topography

1905 'Mapping the Grand Canyon', *MIT Tech. Rev.*, Vol. 7, 1–25.

1907 'The new map of the Yosemite Valley', *Science*, n.s., Vol. 26, 146–8.

1908 'The relation of geology to topography' (with D.W. Johnson), in C.B. Breed and G.L. Hosmer (eds), *Principles and Practice of Surveying*, New York, J. Wiley and Sons, Vol. 2, 246–66.

1912 'Topology, topography, and topometry', *Bul. Am. Geogr. Soc.*, Vol. 44, 334–9.

1915 'The conference on the delineation of physiographic provinces in the United States', *Ann. Assoc. Amer. Geogr.*, Vol. 4, 127–9.

1915 'The survey of Mount Rainier', *The Mountaineer*, Vol. 8, 61–6.

1923 'The story of the Yosemite Valley' (Text on back of topographical map, which stands as 'one of the most beautiful topographic maps ever published'), Yosemite Valley, Yosemite National Park, Mariposa County, California, US Geological Survey.

1926 'Map of Grand Canyon National Park' (with R.T. Evans), *The Military Engineer*, Vol. 18, 188–201.

1928 'The Grand Canyon of the Colorado River' (Text on back of topographical map – an example of 'topographic excellence which [has] never been surpassed'), Bright Angel, Coconino County, Arizona, US Geological Survey.

b. Glaciology and Glacial Geology

1900 'Glacial sculpture of the Big Horn Mountains, Wyoming', *21st Annu. Rep.*, US Geological Survey, Part 2, 167–90.

1904 'The Alps of Montana', *Appalachia*, Vol. 10, 255–76.

1912 'Sketch of Yosemite National Park and an account of the origin of the Yosemite and Hetch Hetchy valley', US Department of the Interior, Washington, DC, 47 pp.

1914 'Mount Rainier and its glaciers, Mount Rainier National Park', US Department of the Interior, Washington, DC, 48 pp.; reprinted (with new title) in E.S. Meany (ed.), 'Glaciers of Mount Rainier', *Mount Rainier: A Record of Exploration*, Macmillan, New York, 1916, ch. 14, 201–40.

1914 'Studying the Yosemite problem', *Bul. Sierra Club*, Vol. 9, 136–47.

1915 'Studies on glacial cirques in the Sierra Nevada', *J. Wash. Acad. Sci.*, Vol. 5, 254–6.

1926 'Kings River Canyon and Yosemite Valley', *Bul. Sierra Club*, Vol. 12, 224–36; reprinted in F. Fryxell (ed.), *François Matthes and the Marks of Time: Yosemite and the High Sierra* (1962), Sierra Club, San Francisco, CA, 85–99.

1928 'The boundary between the Pleistocene and the Recent in the Sierra Nevada of California', *Proceedings Third Pan-Pacific Science Congress*, Tokyo, 1926, 1766–7.

1929 'Multiple glaciation in the Sierra Nevada', *Science*, n.s., Vol. 70, 75–6.

1930 *Geologic History of the Yosemite Valley*, Professional Paper 160, US Geological Survey, 137 pp.

1931 'Mountain glaciers and their work', *Home Geogr. Mon.*, Vol. 1, 37–43.

1932 'Report of the Committee on Glaciers', *Trans. Am. Geophys. Union*, 13th Annual Meeting, 282–7. (From 1932 to 1946 Matthes wrote detailed reports on glaciers, especially on their annual fluctuations, which were published in the *Transactions* of the Union.)

1933 'Geography and geology of the Sierra Nevada', 16th International Geological Congress, United States, 1933, Guidebook 16, 26–40.

1934 'Ablation of snow fields at high altitudes by radiant solar heat', *Trans. Am. Geophys. Union*, 15th Annual Meeting, Part 2, 380–5.

1937 'The geologic history of Mount Whitney', *Bul. Sierra Club*, Vol. 22, 1–18; reprinted in Fryxell, *op. cit.* (1962), 163–83.

1938 'Avalanche sculpture in the Sierra Nevada of California', *Bul. Int. Geod.*

and Geophys. Union, Association of International Hydrographical Science, Vol. 23, 631–7; reprinted in Fryxell, *op. cit.* (1962), 155–62.

1942 'Glaciers', in O.E. Meinzer (ed.), *Physics of the Earth*, Part IX, *Hydrology*, McGraw-Hill, New York, 149–219.

1947 'A geologist's view', in R. Peattie (ed.), *The Sierra Nevada: The Range of Light*, Vanguard Press, New York, ch. 5, 166–214.

1948 'Moraines with ice cores in the Sierra (California)', *Bul. Sierra Club*, Vol. 33, 87–96.

c. Miscellaneous

1910 'Air currents in mountain valleys', *Science*, n.s., Vol. 32, 577–8.

1933 'Cretaceous sediments in Crowley's Ridge, southeastern Missouri', *Bul. Am. Assoc. Petro. Geol.*, Vol. 17, 1003–15.

1934 'Pleistocene changes in the courses of the Mississippi and Ohio Rivers with respect to Crowley's Ridge', *Proc. Geol. Soc. Am.*, 1933, 96–7.

1943 'Max Harrison Demorest (1910–1942)', *Trans. Am. Geophys. Union*, 24th Annual Meeting, Part 1, 12–16.

1946 'The glacial anticyclone theory examined in the light of recent meteorological data from Greenland, Part 1', *Trans. Am. Geophys. Union*, Vol. 27, 324–41.

1947 'Founding of the British Glaciological Society and the *Journal of Glaciology*', *Geogr. Rev.*, Vol. 37, 158–9.

d. Posthumous Works

1950 'The glacial anticyclone theory examined in the light of recent meteorological data from Greenland, Part 2' (with A.D. Belmont), *Trans. Am. Geophys. Union*, Vol. 31, 174–82.

1950 *Sequoia National Park: A Geological Album*, University of California Press, Berkeley, CA, ed. F. Fryxell, 136 pp.

1950 *The Incomparable Valley: A Geologic Interpretation of the Yosemite*, University of California Press, Berkeley, CA, ed. F. Fryxell, 160 pp.; reprinted in 1956.

1960 'A reconnaissance of the geomorphology and glacial geology of the San Joaquin Basin, Sierra Nevada, California', *Professional Paper 329*, US Geological Survey, ed. F. Fryxell, 62 pp.

1962 *François Matthes and the Marks of Time: Yosemite and the High Sierra*, Sierra Club, San Francisco, CA, ed. F. Fryxell, 189 pp.

1965 'Glacial reconnaissance of Sequoia National Park, California', *Professional Paper 504*, US Geological Survey, ed. F. Fryxell, 58 pp.

e. Archival Sources

François Matthes' official papers (field notebooks and photographic album) are in the archives of the US Geological Survey, Federal Building, Denver, Colorado. His personal papers (diaries and some correspondence, old manuscripts and notes) are in the Bancroft Library, University of California, Berkeley, California. Personal mementoes are in the Yosemite Museum.

Fritiof Fryxell (1900–86) was a distinguished teacher of geology at Augustana College, Rock Island, Illinois, 1929–68; a scholar; and literary executor of François Matthes, 1948–65. A former student of Dr Fryxell, William Speth has taught geography at various colleges throughout the United States and published research on the methodology and history of geography. He resides at 1006 E. Third Avenue, Ellensburg, Washington 98926.

Chronology

1874	Born in Amsterdam, the Netherlands, on 16 March, the twin of Gerard Hendrik Matthes, hydrographical engineer
1891	Came to US after education in the Netherlands, Switzerland, and Germany
1895	B.S. *cum laude* in engineering from Massachusetts Institute of Technology
1896	Joined the Topographic Branch of the US Geological Survey as field assistant
1898	Advanced to assistant topographer and designated party chief in charge of surveying the Cloud Peak quadrangle in the Bighorn Mountains of Wyoming
1899	Promoted to topographer
1900	Publication of first scientific work, *Glacial Sculpture of the Bighorn Mountains, Wyoming*
1902	Began mapping the scenically remarkable upper half of the Grand Canyon of the Colorado River, Arizona (now Grand Canyon National Park)
1904	Spent academic year 1904–05 doing postgraduate work at Harvard University in geomorphical studies under W.M. Davis; founder member of the Association of American Geographers
1906	Mapped the San Andreas Fault in northern California for the California State Earthquake Investigation Commission
1907	Completed the exquisite 'Yosemite Special' map at a scale of 1:24 000; promoted to topographical inspector (inspector of maps)
1909	Secretary of the Geological Society of Washington (to 1911)
1910	Placed in charge of mapping Mount Rainier National Park, Washington
1911	Married Edith Lovell Coyle, a native of Washington, DC
1913	Transferred to the Geologic Branch of the US Geological Survey as associate geologist; first geological assignment was to study the origin of the Yosemite Valley; treasurer of the Association of American Geographers (to 1920)
1917	Promoted to geologist

1928	Assigned to investigate the physiographical history of the Mississippi River; promoted to senior geologist
1930	Publication of *The Geologic History of the Yosemite Valley*, which received immediate acclaim as a great classic
1932	President of the Geological Society of Washington; chair of the Committee on Glaciers of the American Geophysical Union (to 1946)
1933	President of the Association of American Geographers; vice-president of the Washington Academy of Sciences
1935	Reassigned to Sequoia National Park, California
1939	Drafted as secretary of the International Association of Scientific Hydrology (to 1948)
1942	Publication of treatise on 'Glaciers' in *Hydrology*, edited by O.E. Meinzer
1945	Undertook critical re-examination of W.H. Hobbs' doctrine of glacial anticyclones over Greenland
1947	Retired from the US Geological Survey after fifty-one years of service
1948	Died in El Cerrito, California, on 21 June

John Muir

1838–1914

Gary L. Peters and Robert P. Larkin

1. Education, Life and Work

John Muir, the eldest son of Daniel and Ann Gilrye Muir, was born on 21 April 1838, in the small Scottish coastal town of Dunbar, near Edinburgh. He was one of eight children.

His father served in the British Army and then sold grain in Dunbar. Daniel Muir was a fanatical Calvinist and played an important, and mostly negative, role in shaping John's attitudes and ideas. Life as a child in the Muir household was grim and austere. There was little laughter or frivolity; John's father ruled the household with the proverbial 'iron fist'. John revealed later that his father thrashed him almost every evening.

John's early school years were spent similarly. School was grim and he hated it. According to John Muir, Scotland's whole educational system was 'founded on leather'. Even when he was quite young, however, people noticed that John had extraordinary talents and interests. As a result of his father's constant goading, John had, by the age of eleven, memorized most of the Old Testament and all of the New Testament. John later commented, 'I could recite the New Testament from the beginning of Matthew to the end of Revelation without a single stop' (cited in Teale, p. 17).

It was only in the outdoors that John Muir found refuge from much of the rigidity of his childhood. He loved to play outdoors with his friends, and, long afterward, at the age of seventy-five, commented in his autobiography, 'With red-blooded playmates, wild as myself, I loved to wander in the fields to hear the birds sing, and along the seashore to gaze and wonder at the shells and seaweeds, eels and crabs. . . . When I was a boy in Scotland I was fond of everything that was wild, and all my life I've been growing fonder and fonder of wild places and wild creatures' (Muir, *The Story of My Boyhood and Youth*, p. 2).

An important milestone in John Muir's life occurred in 1849, when he was eleven. His father decided that the family should move to America and on 19 March 1849 they sailed for New York. They journeyed first to Albany, then to Buffalo, on across the Great Lakes to Milwaukee, and eventually to south-central Wisconsin.

Much of the manual labour involved in clearing the land and developing the farmstead fell on John's shoulders. He often worked as much as seventeen hours a day, seven days a week. Despite this prodigious labour burden, he found time to educate himself in a setting where no formal schooling was available. John taught himself algebra, geometry and trigonometry, and developed an insatiable appetite for reading. His favourite writers included Shakespeare, Milton and Alexander von Humboldt, and he was especially delighted by the geographical exploits of Humboldt.

Juxtaposed with reading, his other passion was inventing and building things. He devised a stream-powered sawmill and made a thermometer, a barometer, and a variety of other odd, but useful, inventions. One of his inventions won first prize at the agricultural fair in Madison, Wisconsin, in 1860. Perhaps more important, however, his inventions caught the eye of the wife of a professor at the University of Wisconsin, who immediately recognized Muir's genius and invited him to become a student at the university.

Muir was at first enthralled by the University of Wisconsin, which at that time had only three buildings and a tuition cost of $32 a year. He moved into a dormitory, set up his own chemistry laboratory in his room, and conducted chemistry and biological experiments. Other students, as well as professors, were amazed at his creativity and inventiveness.

He was especially delighted to work with Professor Ezra Carr, a science professor at the university. Carr taught Muir about the emerging science of glaciology and shared his personal library with Muir. The works of Wordsworth, Emerson, Thoreau, Humboldt and Agassiz were of special interest.

In his junior year Muir decided to leave the university and start 'wandering'. Ironically, he never received his bachelor's degree from the university, but 34 years later he was given an honorary doctorate. After wandering round much of southeast Canada and the north-central USA, working in a variety of industrial jobs, and almost losing his eyesight in an industrial accident, he decided that he would follow in the footsteps of Humboldt and go to South America. 'I might have become a millionaire,' he later said somewhat grandly, 'but I chose to become a tramp' (D. Jones, p. 39).

On 1 September 1867, at twenty-nine years of age, Muir headed south from an area near Indianapolis, Indiana, to the Gulf of Mexico, 'joyful and free, on a thousand-mile walk to the Gulf of Mexico by the wildest, leafiest, and least trodden way I could find . . . with a plant-press on my back, holding a generally southward course, like the birds when they are going from summer to winter' (Muir, *A Thousand-Mile Walk to the Gulf*, p. 1).

Although only two years after the American Civil War was a harrowing time for a 'northerner' to travel in the South, Muir eventually made it to Cedar Keys in Florida. An account of this journey and the environmental observations made by Muir were eventually published in his book *A Thousand-Mile Walk to the Gulf*.

Muir's original intention of following in the footsteps of Humboldt was thwarted when he contracted malaria and was forced by poor health to abandon his South American journey. Instead, he decided to head west to California, where in 1868 he arrived in San Francisco and then went on to the Yosemite area of the Sierras. While in Yosemite he became fascinated by the region's geology and ecology. The unity of nature focused his thoughts and solidified his belief that 'when we try to pick out anything by itself, we find it hitched to everything else in the universe' (Wolfe, p. 124).

While in Yosemite he remembered Dr Carr's lessons on glaciology and studied the evidence of glacial action in the Yosemite region. As a result, he published several articles on the region's glacial evolution. They were not well received by many of the scientific community. Josiah Whitney and Clarence Starr King, leaders in the emerging field of geology and geomorphology, looked scornfully at Muir's ideas and ridiculed him as a 'mere sheepherder', an 'ignoramus'.

On the other hand, he was supported by several notable scientists and naturalists of the time. Louis Agassiz said of Muir, 'he is the first man who has any adequate conception of glacial action' (Clarke, p. 108). Professor Joseph LeConte of Berkeley and John Tyndall of the British Royal Institution also supported Muir's ideas about glacial action in Yosemite. Muir also had the respect of Ralph Waldo Emerson. After a meeting with Emerson at Yosemite, Emerson added Muir's name to 'My Men', a short list of acquaintances such as Carlyle, Wordsworth and Coleridge whom he found most 'excellent' (Clarke, p. 93). Muir also had a visit from John Daniel Runkle, president of the Massachusetts Institute of Technology, who offered him a position on the MIT faculty.

During the early 1870s Muir continued his studies of the Sierras but broadened them somewhat to include more ecological variables. He was far better equipped to observe and interpret than he had been during his first summer because many able scientists of the day visited him (John Tyndall, Asa Gray, John Torrey) and provided him with an on-site postgraduate faculty. He now wanted to elucidate what he called 'the mystery of harmony'. These observations, dealing with life-forms in relation to altitude and other phenomena, were the first systematic observations of life zones in the Sierra Nevada and among the earliest made anywhere in the world. Muir's fascination with the work of Humboldt is evident in these 'Studies in the Sierra', which were published in the *Overland Monthly*.

During the mid-1870s Muir turned his attention to the giant sequoias. In a paper for the American Association for the Advancement of Science, he concluded that the big trees, rather than seeking wet areas, actually created such areas by capturing water that would normally come roaring downslope in destructive floods. He also became concerned about the practices of lumber companies and sheepmen and their influence on the environment. He presented a paper on the urgency of protecting the Sierra forest before the American Association for the Advancement of Science at its twenty-fifth annual meeting, and it was published in the Association's *Proceedings*.

Muir's articles on redwoods and other pieces on ecological issues, published in a variety of popular journals such as *Harper's Magazine* and *Scribner's Monthly*, were instrumental in establishing a national reputation for him. They also were a means of meeting many influential friends who later were important in furthering his efforts at wilderness preservation. Throughout the later 1870s he travelled in the West (Oregon, Washington, Utah). It was at that time that he was attracted to the daughter of a Polish scientist and physician. Her name was Louie Wanda Strentzel, and he courted her during the summer of 1878. At this time he also met a missionary who had recently been in Alaska and who told Muir about the many glaciers there.

Little more than a decade had passed since the USA had purchased Alaska from the Tsar of Russia and little was known about it except that it was very cold and dangerous. The lure of Alaska proved to be too great, and Muir postponed his engagement to Louie Wanda and left San Francisco for Alaska in May 1879. It was to be the first of five trips to Alaska. On this first trip he sailed into a small bay that did not appear even on the charts Captain Vancouver had made ninety years earlier. This bay, later named Glacier Bay because of its tidewater glaciers, enthralled Muir, who commented, 'Never before this had I been embosomed in scenery so hopelessly beyond description' (Clarke, p. 187).

The first glacier he encountered he later named Geike, after a famous Scottish geologist. Just beyond the point where the northwest fork of Glacier Bay begins, Muir encountered a large glacier a mile and one-half wide and in places 700 feet high. This was the first recorded sighting of the glacier that later was to bear Muir's name.

During another excursion, near the site of the present-day city of Juneau, Muir met two prospectors, Joe Harris and Joe Juneau, and mentioned to them that the gravel bed of a nearby stream might contain gold. Before long Juneau and Harris found rich quartz deposits that led to the historic Alaska gold stampede (Wolfe, p. 217).

Upon returning from this Alaska trip he married Louie Wanda on 14 April 1880, and started what eventually proved to be a highly profitable farming enterprise near Martinez, California. His botanical training was a great asset in his experiments with different varieties of fruit and he became a very successful farmer and fruit merchant. The money began quite literally to 'pile up'.

Although the 1880s and 1890s brought financial prosperity and independence for Muir, he continued his writing and advocacy for environmental causes. He made additional trips to Alaska, and during one trip through the Bering Strait he observed the landscape on both the Alaskan and Siberian coastlines. From that observation he concluded that the Asian and North American continents had at one time been connected by a land-bridge. Though generally accepted today, such an idea had not even been considered in the 1880s.

The 1890s was a decade of activism for Muir. In early 1890 he wrote two articles about Yosemite that were published by *Century* in August and September. They called for the preservation of the Yosemite region and made the case that if citizens wanted to keep the country they loved habitable they would have to fight for its preservation. By this time Muir had published some sixty-five articles and had a considerable number of followers. His strong advocacy for forest preservation gave rise to the often-repeated statement that Muir was 'the father of the American National Forests' (Clarke, p. 265). In 1892 he met informally with friends and drafted the articles of incorporation for the Sierra Club, which was to become one of the most influential environmental organizations.

Muir believed that the only way to save the wilderness was for people to come to love it. He was encouraged by friends to change his writing genre from articles to books. Thus, in 1894 Muir's first book, *The Mountains of California*, was published. It was a biogeographical study of the Sierra Nevada and was an instant success – the first printing sold out within weeks of publication. Even the British biologist Sir Joseph Hooker wrote to Muir, 'I do not know when I have read anything I have enjoyed more. It has brought California back to my memory with redoubled interest and more than redoubled knowledge' (Clarke, p. 274).

After the turn of the century Muir continued his environmental efforts. He made friends with a variety of influential people, including President Theodore Roosevelt, and continued his book writing. In 1901 his second book, *Our National Parks*, was published. In 1903, at sixty-five years of age, he embarked on a year-long, worldwide tour that took him through Europe and Russia, Manchuria, India, Egypt, Australia, New Zealand, the Philippines, Japan and Hawaii. It would be remarkable, even today, to embark on such a grand tour, but that it was undertaken by Muir at age sixty-five is a testimony to his vigour and commitment to see 'God's glorious wilderness' (Muir, letter to William Keith, 28 March 1904). This grand tour served as a means of renewing Muir's commitment to nature and wilderness. He needed those trips away from the farm to recharge his enthusiasm for nature (Cohen, 1984). Colin Hall, in a recent article about Muir's geographical work in New Zealand during the grand tour, suggested that 'Muir's travels through New Zealand may therefore be seen as an essential part of Muir's preparation for the efforts required to preserve America's wilderness heritage upon his return to the United States' (Hall, p. 102).

Following this world tour, Muir actively campaigned for the preservation of Hetch-Hetchy Valley. For Muir, a plan to drown Hetch-Hetchy Valley, north-east of Yosemite Valley, was an outrage to all Americans. Although Hetch-Hetchy had been a national park since 1890, the increasing demand for water by the growing city of San Francisco put utilitarian interests above those of conservation and preservation. The effort by Muir and his followers to preserve this region is the first example of a nationwide wilderness preservation campaign anywhere in the world (Hall, p. 100). This campaign acted as a catalyst for the creation of the US National Park Service (Jones, 1965). Unfortunately, Muir lost the battle, and a dam was built across Hetch-Hetchy Valley.

Muir continued his writing, and in 1911 his third book, *My First Summer in the Sierra*, was published. It was also the year that, at age seventy-three, he embarked on his long-deferred South American expedition. He would finally get a chance to walk in the footsteps of Humboldt and fulfil his boyhood dream. His fourth book, *The Story of My Boyhood and Youth*, was published in 1913, the same year that he received an honorary doctorate from the University of California at Berkeley. It was his third such degree. Because he was a college drop-out, lacking even a BA, he was never given an academic degree, but 'on the other hand, he was a fellow of the AAAS, and his contributions to geology, botany, geography, and ecology were more significant than many which have been rewarded by doctorates' (Clarke, p. 316).

Muir continued working 12- to 16-hour days, as he was remodelling his Martinez home and working on his fifth book, *Travels in Alaska*. In the latter part of 1914 he developed pneumonia and was hospitalized in Los Angeles, where on Christmas Eve, with proof-sheets of *Travels in Alaska* spread out on his bed, he died.

2. Scientific Ideas and Geographical Thought

John Muir can easily be considered a *sub rosa* geographer. Though his appreciation of place and landscape would probably be sufficient justification, his major scientific contributions to geography would include glacial studies, an ecological approach to conservation, and regional studies of the western USA, especially California and Alaska. Though most of his major contributions have already been mentioned, they are summarized here again.

He contributed much to our early geographical understanding of glaciers, both as a process that shaped alpine landscapes, borne out by his personal observations in Yosemite Valley and later in Alaska, and as evidence that past geographies must have differed from what we can observe today. More than one hundred years ago he was the first to note that there had been a land-bridge between Eurasia and North America during some past, and obviously very different, period of geological time. Though it was left for others to chronicle in subsequent studies the chronology and details of Pleistocene geography, he had the imagination to initiate the idea.

Ecological relationships fascinated Muir throughout his studies of the Sierra

Nevada, and he forces us to realize the degree to which ecosystems are complex webs of interactions in which the parts are less than their sum as viewed in the whole. His sense of interconnectedness also led him to see that human actions could lead to dire consequences for ecosystems. The grazing of cattle and sheep accelerated erosion, disturbing the delicate natural balance of those mountain landscapes that had become so much a part of him. He spoke out on this issue, and in the process introduced an activist approach to conservation that still continues, though in recent decades it has ebbed and flowed with the state of the economy and perceived ecological urgency. He led the fight over Hetch-Hetchy, and though he lost that battle he gave prominence to the trade-offs that must be made between natural environments and the needs of industrial societies. The Sierra Club has survived as part of Muir's legacy, and its continued strength is a tribute to the depth of his perception about human/environment conflicts.

Anyone who has read Muir's books, whether on the Sierra Nevada that he loved so much, on Alaska, or on some other region, cannot help but be struck by his ability to observe and interpret the landscapes through which he travelled. It is that combination of observation and interpretation that endears him to those who believe that good regional geography is hard to beat. After all, regional geography extends his ideas of the connectedness of things in natural ecosystems to include those things added by human beings in their never-ending commitment to 'progress', with all that that term connotes.

3. Influence and Spread of Ideas

Had he been born a century later, his life and work would have been much different. Because he never graduated from college, he lacked the 'credentials' that would probably have helped him find more and quicker acceptance of his work. Though we now recognize how important his works on glaciation really were, they were not widely accepted without considerable difficulty in the scientific community of the late nineteenth century. He did manage, finally, to find support for many of his ideas, but in retrospect the very nature of scientific advancement has left some of his earlier ideas behind, though not without credit.

For geographers, however, he continues to be of importance because he wrote so elegantly and persuasively about the natural world that he observed, whether in the Yosemite Valley, the High Sierra or Glacier Bay. His words often re-create that 'sense of place' that is so elusive, yet so important. His writings appeared in scientific journals, popular magazines, and books, though it is the last-named that are most likely to be read today, especially those that were written about the Sierra Nevada. Today his work is perpetuated in publications from the Sierra Club, where beautifully photographed books are accompanied by timeless quotations from his writings.

His legacy for geography, then, is found in the combination of relationships that he was able to observe, record and interpret. In modern terms, he believed in 'general systems theory' as a way of organizing his observations and using them to explain and predict relationships in ecosystems, especially mountain ecosystems. From such thinking, he detected the problems that were being created by the expansion of human activities, especially the transhumance that brought 'hooved locusts' to the mountains in summer and the growth of cities that ultimately brought people to the mountains in search of water. Hetch-Hetchy remains a symbol for conservationists. Wherever people gather together to talk about national parks, about preservation of the wild, or about the continuing clashes between environmentalists and industrialists, the spirit of John Muir will be present, and his name will be recognized by all those present.

Bibliography and Sources

1. REFERENCES ON JOHN MUIR

Clarke, James M., *The Life and Adventures of John Muir*, San Francisco, CA: Sierra Club Books, 1980.

Cohen, Michael P., *The Pathless Way: John Muir and American Wilderness*, Madison, WI: University of Wisconsin Press, 1984.

Fox, Stephen, *The American Conservation Movement: John Muir and His Legacy*, Madison, WI: University of Wisconsin Press, 1985.

Hall, Colin Michael, 'John Muir in New Zealand', *New Zealand Geographer*, Oct. 1987, pp. 99–103.

Jones, Dewitt and T.H. Watkins, *John Muir's America*, Portland, OR: Graphic Arts Center Publishing Company, 1976.

Jones, H.R., *John Muir and the Sierra Club: The Battle for Yosemite*, San Francisco, CA: Sierra Club Books, 1965.

Kimes, William F. and Maymie B. Kimes, *John Muir: A Reading Bibliography*, Palo Alto, CA: William P. Wrendon, 1978.

Teale, Edwin W., *The Wilderness World of John Muir*, Boston, MA: Houghton Mifflin Company, 1954.

Turner, Frederick, *Rediscovering America: John Muir in His Time and Ours*, San Francisco, CA: Sierra Club Books, 1985.

Wolfe, Linnie M., *The Life of John Muir*, Madison, WI: University of Wisconsin Press, 1945.

2. SELECTIVE BIBLIOGRAPHY OF WORKS BY JOHN MUIR

1872 'Living glaciers of California', *Overland*, Vol. 9 (Dec.), 547–9.

—— 'Rambles of a botanist among the plants and climates of California', *Old and New*, Vol. 5 (June), 767–72.

1873 'A geologist's winter walk', *Overland*, Vol. 10 (Apr.), 355–8.

—— 'On the effects of the earthquake of March 26, 1872, in the Yosemite Valley', *Proc. Boston Soc. Nat. Hist.*, Vol. 15, 185–7.

1874 'Formation of soils', *Overland*, Vol. 13 (Dec.), 530–40.

—— 'On the mountain sculpture in the Sierra Nevada, and the method of glacial erosion', *Am. J. Sci.*, Vol. 7 (May), 515–16.

—— 'Post-glacial denudation', *Overland*, Vol. 13 (Nov.), 393–402.

1875 'Flood-storm in the Sierra', *Overland*, Vol. 14 (June), 489–96.

—— 'The formation of mountains in the Sierra', *Proc. Am. Assoc. Adv. Sci.*, Vol. 23 (Aug.), Part B, 49–64.

1877 'On the Post-glacial history of the Sequoia Gigantea', *Proc. Am. Assoc. Adv. Sci.*, Vol. 25, Part B, 242–53.

1878 'Douglass Squirrel of California', *Scribner's Monthly*, Vol. 17 (Dec.), 260–6.

—— 'Humming-bird of the California water-falls', *Scribner's Monthly*, Vol. 15 (Feb.), 545–54.

—— 'New Sequoia forests of California', *Harper's Magazine*, Vol. 57 (Nov.), 813–27.

1879 'The glacier meadows of the Sierra', *Scribner's Monthly*, Vol. 17 (Feb.), 478–83.

1880 'Ancient glaciers of Sierra', *Californian*, Vol. 2 (Dec.), 550–7.

1881 'Coniferous forests of the Sierra Nevada', *Scribner's Monthly*, Vol. 22 (Sept.), 710–23, 921–31.

1882 'The bee-pastures of California', *Century*, Vol. 2, 222–9 and 388–96.

1890 'Features of the proposed Yosemite National Park', *Century*, Vol. 40 (Sept.), 656–67.

—— 'The treasures of the Yosemite', *Century*, Vol. 40 (Aug.), 483–500.

1893 'Alaska', *Am. Geol.*, Vol. 11 (May), 287–99.

1894 *The Mountains of California*, New York: Century Press.

1895 'The discovery of Glacier Bay', *Century*, Vol. 50 (June), 234–47.

1897 'An adventure with a dog and a glacier', *Century*, Vol. 54 (Sept.), 769–76.

—— 'The Alaska trip', *Century*, Vol. 54 (Aug.), 513–26.

—— 'The American forests', *Atlantic*, Vol. 80 (Aug.), 145–57.

—— 'The national parks and forest reservations', *Harper's Weekly*, Vol. 41 (5 June), 563–7.

1899 'The Yosemite National Park', *Atlantic*, Vol. 84 (Aug.), 145–52.

1901 *Our national parks*, Boston, MA: Houghton Mifflin.

—— 'The fountains and streams of the Yosemite National Park', *Atlantic*, Vol. 87 (Apr.), 556–65.

1907 'The Tuolumne Yosemite in danger', *Outlook*, Vol. 87 (Nov.), 486–9.

1908 'The Hetch Hetchy Valley', *Bul. Sierra Club*, Vol. 6 (Jan.), 211–20.

1909 'The endangered valley: The Hetch Hetchy valley in the Yosemite National Park', *Century*, Vol. 77 (Jan.), 464–9.

1910 'The Hetch-Hetchy Valley, a national question', *Am. For.*, Vol. 16 (May), 262–9.

1911 *Edward Henry Harriman*. New York: Doubleday Page.

—— *My first summer in the Sierra*, Boston, MA: Houghton Mifflin.

1912 'The plunge into the wilderness', *Atlantic*, Vol. 110 (Dec.), 813–25.

1913 *The story of my boyhood and youth*, Boston, MA: Houghton Mifflin.

1915 *Travels in Alaska*, Boston, MA: Houghton Mifflin.

1917 *A Thousand-Mile Walk to the Gulf*, edited by William Frederic Bade, Boston, MA: Houghton Mifflin.

Gary L. Peters is Professor of Geography at California State University, Long Beach, California. Robert P. Larkin is Professor of Geography at the University of Colorado, Colorado Springs, Colorado.

Chronology

1838	Born in Dunbar, Scotland, 21 April
1849	Emigration of Muir family to America
1857	Muir family moves to Hickory Hill farm, Wisconsin
1861–63	Undergraduate at the University of Wisconsin, Madison
1867	Muir's thousand-mile walk to the Gulf
1868	Muir's first visit to Yosemite
1871	Visit to Yosemite by Ralph Waldo Emerson; Muir's first study of glaciers written
1872	Begins writing for the *Overland Monthly*
1879	Engaged to Louie Wanda Strentzel; leaves on first Alaskan expedition
1880	Marries Louie Strentzel, 14 April
1881	Daughter, Annie Wanda, born; second trip to Alaska and discovery of Bering Strait land-bridge
1890	Begins writing for the *Century*
1892	Sierra Club founded by Muir
1894	*The Mountains of California*, Muir's first book, published
1901	*Our National Parks* published
1903	Muir meets President Theodore Roosevelt; embarks on worldwide journey
1905	Death of Muir's wife
1908	Creation of Muir Woods National Monument
1911	*My First Summer in the Sierra* published; Muir departs on long-deferred South American expedition
1913	*The Story of My Boyhood and Youth* published
1914	Muir dies in Los Angeles, 24 December

Stanisław Pawłowski

1882–1940

Witold Wilczynski

In the history of Polish geography and national culture of the last two centuries there was only one short period when the circumstances for free development existed. That was the period between World War I and World War II, when Poland became an independent state. Earlier, in those parts of Poland which fell within the Russian Empire and the Kingdom of Prussia, a Polish educational system did not exist. For Poles a scientific career was possible only in the framework of research carried out at German or Russian institutions. Censorship restricted the publication of Polish scientific works, and few Polish scientists were able to develop their own concepts (see 'Wacław Nałkowski, 1851–1911', *Geographers*, Vol. 13). Only in the southern part of partitioned Poland, which was on the borders of the Austrian Empire, was a broader range of political freedom allowed for the development of Polish science. Two centres of Polish culture exerted an influence from there over all Polish territory. These were Kraków with its very old Jagiellonian University and Lwów (Lemberg), where the John Casimir University was established. In those two centres a Polish scientific faculty was trained, and scientists from Kraków and Lwów assumed a major role in Polish scientific development, after 1918, in independent Poland. Stanisław Pawłowski was for most of his scientific career the director of the Geographical Institute at Adam Mickiewicz University in Poznań; however, he came from the Lwów school of geography, created by Eugeniusz Romer, one of the founding fathers of Polish geography.

1. Education, Life and Work

Stanisław Pawłowski was born on 16 March 1882 in the small Carpathian settlement of Dembowiec, in Galicia. His family had been weavers for a number of generations. After primary and grammar school education, the latter in Sambor, he joined a secret patriotic organization; this forced him to move to the gymnasium in Jasło. In 1902 he commenced university studies in Vienna, where he attended the lectures of Albrecht Penck, Edward Suess and Karl Uhlig. He continued his studies in Lwów, where he met the historian Bronisław Dembinski and Eugeniusz Romer. His fundamental interest was history, to which he made a meaningful contribution with a dissertation on Constantine the Great. His return to geography was due to the attractive and suggestive lectures of Romer.

Pawłowski began his pedagogical work as a teacher in the gymnasium in Stanisławów (Ivano-Frankovsk) in 1906. In 1908 he completed his studies and became the close co-worker of his master, Eugeniusz Romer. In 1910 he defended a doctoral thesis on 'The Temperature of Water in Galician Rivers' and was awarded a Ph.D. degree by John Casimir University in Lwów. For eight years he stayed in Lwów as a lecturer in geography at the Academy of Commerce and a secondary schoolteacher. At the same time he undertook morphological fieldwork in the eastern part of the Carpathian Mountains and the Little Polish highlands. He was particularly interested in studies on glaciation in Carpathia. His first textbooks for secondary schools were published, as was his *Geography of Poland* (1917), a regional synthesis of the historical territory of Poland. In 1918 he was appointed professor of geography at John Casimir University. In 1919 he began to organize a geographical institute in the new Polish university in Poznań. The location of Poznań on the then western fringe of Poland brought him up against the geographical and political problems of that region. Those territories were open to Polish science and Pawłowski began to address such matters.

The removal of Pawłowski from the Little Polish highlands to the Pleistocene lowlands of Greater Poland brought about diversification in his research. He gave particular attention to the morphology of Greater Poland and Pomerania as well as the Baltic seashore. He had examined the adequacy of earlier German ideas concerning postglacial morphology, and he showed specific morphological characteristics of those lands on the basis of concepts derived from Vienna and the Romer school. He had become the authority on the glacial morphology of Poland and Pomerania. His division of the Polish lowlands into morphological regions is still accepted. His interest in highlands did not disappear when he relocated to the lowlands. During his holidays, which he took most frequently in his native Carpathia, he carried out further morphological observations. As a result, he prepared a synthesis of glaciation in the Carpathian Mountains, which he presented at the Fourteenth International Geographical Union Congress in Warsaw in 1934.

Like his teacher, Romer, and the majority of Polish geographers, he read widely in many branches of geography. During his stay in Lwów he published the first Polish synthesis of human geography in which he wrote of the development of human geography from early times to Karl Ritter, Friedrich Ratzel, the French regionalists, and the Polish geographer Wacław Nałkowski, whose holistic views he shared. During field trips Pawłowski never limited his observations to morphological phenomena. In the north-western part of Poland, which was the Polish-German frontier, he had perceived the problems connected with the Slavonic past and the contemporary geographical-political question, and he championed Polish culture in the areas threatened by German influence for many centuries. He promoted the 'Slavia Occidentalis' edition of the *Atlas of Geographical Names of Western-Slavonic Countries* (only two volumes were edited before 1939). In his research on rural settlement in Greater Poland and Pomerania he had demonstrated a reduction of landed property in Polish hands and its concentration in the hands of a German minority. As a geographer, he participated in discussions on Polish boundaries and he published some works concerning the plebiscite areas on the eastern bank of the lower Vistula, the significance of access of Poland to the sea, and the problem of the Gdańsk Independent City. He welcomed political geography as a field which would help fashion people's international consciousness, and he was optimistic in the matter of international co-operation.

Pawłowski's geographical publications have a strong didactic emphasis. He considered that a large opportunity for geographical application resided in education, and the enlargement of social consciousness. As a promoter and chairman of the Association of Polish Geography Teachers established in 1922, he protected and helped teachers, and defended geography in the school curricula. In 1930 he arranged the first Faculty of Education in Poland. He prepared three editorial series for teachers and youth. They included 'Lektura Geograficzna' (Geographical Reading Matter), 'Dookola Swiata' (Round the World), and 'Biblioteka Geograficzno-dydaktyczna' (Geographical Education Library). The publishing house of Ksiaznica-Atlas in Lwów published more than 70 volumes in these series in the 1930s.

A significant part of Pawłowski's activity seems to have been his international connections. He was a good friend of Emmanuel de Martonne, and prepared a Polish edition of Martonne's *Abrégé de géographie physique*. Pawłowski had travelled and studied abroad: North Africa (1924), Wales and France (1928), Ethiopia (1935), Norway (1937), and Indonesia and India (1938). At the Geographical Institute in Poznań, Pawłowski created a library and arranged the scientific editorial series entitled 'Geographical Investigations in North-Western Poland' (later 'Geographical Investigations'). He organized numerous conferences and symposia concerning western Poland. Pawłowski functioned successfully as chairman of the Friends of Science Association, the Geographical Society, the Association of Polish Geography Teachers, and Rector of Adam Mickiewicz University, among other tasks. In 1934 he was secretary-general of the Organizing Committee of the Fourteenth International Geographical Congress in Warsaw. Moreover, in 1938, during the Fifteenth International Congress in Amsterdam, he was elected a vice-president of the International Geographical Union.

In Pawłowski were fused the characteristics of outstanding scientist, splendid teacher and successful organizer. His lectures attracted ever more listeners. For his students he was rigorous, maintaining strict requirements, though always allowing freedom of expression. He provided support for any person who showed geographical interest. There was a strong and friendly relation between Pawłowski, his students and geography teachers. His seeming inaccessibility resulted rather from his shyness. On field trips he was always merry, talkative and inclined to sing! The spirit of Pawłowski was a fusion of discipline and a sense of organization with the originality, imagination and fantasy derived from Poland's eastern borderland.

Such characteristics allowed him, with a Sarmatian spirit, to awaken among Poles resistance in the face of Germanization, which had endangered Poland for centuries. He sought to introduce effectively the science of geography – 'which was to join Polish people with the land' – in the territories where the Polish state had come into existence ten centuries before. In considering the educational meaning of geography, Pawłowski had written: 'It is a universally shared view, that geography awakes a love towards the native land.' That was the message of Pawłowski's geography. He had undertaken this task and sealed it with his own blood. In October 1939 he was arrested by the Gestapo, who accused him of 'hostility towards Germans'. He was shot dead in the Seventh Fort of Poznań in January 1940, and his remains were buried together with those of 1700 other Poles in the common grave at the Wypałanki forest near Poznań.

In 1965 the Polish Geographical Society organized a symposium devoted to Pawłowski on the occasion of the twenty-fifth anniversary of his death. Stanisław Leszczycki acknowledged the great loss Polish geography suffered with Pawłowski's death: 'There in Poznań did not survive the main representative of geography, the powerful Poznańian oak . . . one of the most powerful pillars of polish geography, which with a steel will, with the highest devotion highlighted Polish geography on the international forum.'

2. Scientific Ideas and Geographical Thought

Pawłowski's scientific output includes more than 330 items, reflecting his wide-ranging interests. Today there no longer exist geographers whose range of interests are comparable with the scope of Pawłowski's thought. The typical contemporary geographer is a specialist. Postwar Polish geography, even if tied to the ideology of Marxism-Leninism, remains strongly imbued with the scientific spirit. World War II was not only a turning-point in the political and social history of Poland; it was also a critical time for Polish geography since a generation of senior Polish geographers was exterminated and a new ideology strongly influencing scientific methodology was imposed.

Stanisław Pawłowski was not only a geomorphologist, hydrologist, settlement geographer and political geographer. He was above all a geographer, undertaking specialist research in seemingly distant fields of geography, though never forgetting the unity of the field. He wrote methodological dissertations and textbooks based on the regional idea. He was interested in all phenomena which constituted the geographical environment. He had a deep humanistic knowledge and an excellent capacity with regard to pedagogical, psychological, and philosophical problems. In contemporary literature he is known essentially as a geomorphologist.

His interests were mainly influenced by his native Carpathian landscapes. His studies in Vienna and in Lwów stirred his emotional, patriotic, and aesthetic excitement. His knowledge of geology further enabled him to understand the landscapes among which he had grown up and which had earlier been only the object of admiration. It was sufficient stimulus for his lifelong engagement in geographical research. His first works concerned hydrography and the palaeogeography of river networks in Galicia. In his studies Davisian concepts and terminology were used; he was one of the pioneers of research carried out in Davisian terms. His later works were connected with the problems of the origin of landscapes and the evolution of morphology in the Carpathian Mountains and in neighbouring areas. He treated the investigation of alluvial fans and river terraces as a key to the understanding of Carpathian morphology.

Together with Romer, or independently, he had explained numerous problems connected with the asymmetry of valleys, the influence of rock varieties on morphology, the existence of genetically different valleys and gorges, and the shifting of watersheds. Based on the antecedent and epigenetic theory, he had explained the origin of the Dunajec River gorge cutting the Pieniny Rocky Belt. In 1916 he synthesized what was known about the morphology of the outer Carpathians. He was an adherent of the view that most forms had been created at the end of the Tertiary period. He was sceptical, however, about the notion of dissected Tertiary plateaux, as it was not substantiated by empirical evidence. Pawłowski is, moreover, an author of the concept which explains the origins of the Podole escarpment. Its existence and the breakdown in long profiles of rivers on its line are, according to Pawłowski, mainly due to the contemporary upward movement of the so-called Metacarpathian Block. This view has been verified by neotectonic measurements.

Pawłowski is one of the main creators of contemporary theory concerning the development of the glacial morphology of the Carpathian Mountains. He had treated cirques and U-shaped valleys as evidence for glaciation and he rejected theories of their tectonic origins. He presented his synthesis of Carpathian glacial morphology to the Fourteenth IGU Congress in Warsaw. There he emphasized the unique character of glacial forms in the outer ridges of the Carpathians and synchronized the Carpathian glacial periods with those of Scandinavia. His analysis of the glacial morphology in Greater Poland and Pomerania is perhaps his greatest contribution. There he had noticed the full range of genetic morphological landscapes in accordance with the kind of glacial sediment and subglacial erosional form. He explained some details in the contemporary relief of Greater Poland by the reconstruction of glacial activity on a Tertiary surface. The next area of his morphological investigation was the Baltic coast; he was the first Polish geographer to explain the origins of Baltic haffs and bars. He examined the previous German concepts on the basis of detailed fieldwork and investigations of the Gdańsk Bay bottom. He introduced some Polish or local Kashub terms to the morphological dictionary, denoting the main morphological phenomena of the seashore. In his morphological research he made use of petrological and stratigraphical analyses.

In relation to morphology Pawłowski was an adherent of analytical, empirical methods. He realized, however, that empirical investigation based on inductive methods is not alone sufficient. Pawłowski declared the methodology of induction insufficient, and acknowledged the need for the introduction of Wundt's interpretative-deductive method: this was prior to the publication of Karl Popper's *Logic der Forschung*. Pawłowski's research carried with it a deep methodological substantiation. The most essential value of Davis's theory for him was not in the scheme of erosion cycles, but in the methods of scientific research. The value of Davisian theory, according to Pawłowski, was to be found in the application of some elements of deductive thinking, which supplement inductive methods. Pawłowski's view on geomorphology is quite modern, since he treated it in a dynamic manner and acknowledged the existence of compound morphogenetic complexes.

Engagement in empirical morphological studies as well as in methodological considerations did not embarrass Pawłowski in research concerning seemingly distant spheres such as population distribution and settlement geography. In 1928 he had commenced research on rural settlement in Greater Poland. He had introduced to Polish settlement geography the new approach connected with Albert Demangeon's concepts and studies carried out by the IGU Commission on Rural Settlement. As in geomorphology, he had emphasized the importance of fieldwork and the role played by settlement in landscape. He established the typology of settlements based on physiognomical criteria. During the Fourteenth Geographical Congress, he and Jozef Czekałski offered a synthetic study of rural settlement in Poland. The explanation of the distribution of rural settlements in Poland is still elusive, since the structure of rural settlement in Poland is complex.

The central interest of Pawłowski was not the relief or the settlement alone, but the whole of the geographical landscape. Since for its understanding the genetic explanation of morphology, as well as the knowledge of historical development of the settlement process and land-use patterns, is necessary, Pawłowski could not ignore these problems. He wanted to reach the level of understanding whereby every landscape's phenomena could be clearly substantiated on the basis of logic, natural laws and history. He realized that natural laws and social phenomena are quite different, and he devoted much energy to the development of methodology. His methodological works are the most original part of his very abundant scientific output. With regard to the understanding of geography, Pawłowski seems to be the faithful disciple of Wacław Nałkowski, the creator of geography as an interdisciplinary, synthesizing field: 'Geography defends us against the penetration of details and specialities. It sets us free from theory and fiction and brings us towards reality.'

Pawłowski did not neglect the connections of detailed studies with the larger and general geographical tasks. Working sometimes with abstract and esoteric 'relations' and 'structures', he never lost sight of reality, the geographical landscape. The notion of landscape or geographical environment he understood in humanistic terms. The unity of the world is to be found, he claimed, in the Earth viewed as a harmonious whole, whose unity is the product of human perception. Of the unity of the world he wrote: 'Physical and human phenomena which display themselves on the stage of the Earth's surface, occur always side by side, always in mutual relation . . . Physical and psychic processes are in co-operation in the frame of the global whole in spite of the fact that these connections are not easy to discover and to substantiate.'

That fragment of Pawłowski's writings seems to be the best evidence of his conviction of the need for the unification of geography. He had marked out two complementary stages in geographical research. The first is represented by general geography, whose aim is the description and discovery of some hidden connections that exist between particular elements of the geographical environment with the use of inductive methods. Generalization is, then, the central task of general geography. The other stage of geographical study is, however, represented by regional geography. We are moving from the sphere of abstract connections typical for general geography to the sphere of perceived and experienced reality. The generalized approach is replaced by that of the individual. The aim is to show the individuality or identity of spatial fragments of geographical environment which arise from the differences in substantial and phenomenal characteristics in the sphere of nature as well as human affairs. The inductive method used in the first stage of geographical study is then supplemented with deduction and interpretation. The strictly scientific explanation is enriched by a humanistic approach.

There are strong analogies between this concept and the postwar ideas of William Kirk concerning phenomenal and behavioural environments. According to Pawłowski, both approaches serve the same aim – the recognition and understanding of the geographical environment. Both are the necessary components of cognitive processes. Pawłowski's stance is thus different from the better-known Hettnerian idea of methodological dualism. In some postwar comments on Pawłowski's works, it was emphasized that he was an adherent of experimental and quantitative methods. However, it is hardly possible to consider Pawłowski a forerunner of postwar 'quantitative geography'. According to Pawłowski, each geographical phenomena is of a compound nature and is distributed over large spaces. He wrote: 'It is difficult to reduce the great laboratory of nature to the form of any small geographical one.' The geographical task is, then, to put these phenomena together with the use of maps and to investigate them with 'systems'.

Pawłowski was conservative with respect to new lines of work and expressed reservations with regard to the application of quantitative methods. He found regional geography to be the best method for disciplinal integration. He was opposed to the German geographers, who wanted to retard the process of specialization by developing general geography. The concept of *Allgemeine Geographie* seemed to Pawłowski a formal measure devoid of creative value. The main pillars of his geographical ideas were the concept of a geographical region as a fundamental research unit, and the concept of landscape treated in a synthetic, humanistic manner. Pawłowski wrote:

> The main task of geography is the landscape synthesis, and its aim . . . is the understanding of landscapes. The specific geographical point of view makes geography the unified field and takes it out from the precipice of encyclopaedism . . . Geography is the lens concentrating and joining and at the same time giving the broad and unified vision of the world . . . Geography builds far bridges, enlightens the huge world's horizons. It is the enemy of one-sided, exclusive views. It is not a building composed of many various elements, but its characteristic is rather a monolithic construction.

It is well to emphasize that Pawłowski had published his synthetic and anthropocentric concept of geography in 1918, prior to the time when some of the works of Brunhes and Vidal de la Blache appeared. In explaining human geography, Pawłowski adopted a sociological approach similar to that of some of his French colleagues. Geographical phenomena are, as he wrote, 'the effect of natural laws as well as of the other factor, that is, human free will'. Both of these groups of factors produce the geographical landscape and its physiognomy.

Pawłowski's devotion to the idea of regional geography and the unity of geography derived from his belief in the worth of this science and its educational value. Pawłowski recognized the main problems of education and found that geography could effectively influence people's consciousness. Pawłowski's geography was in the field, the beginning place for education. The geographical curriculum of Pawłowski with regard to substance as well as method was well adjusted to the psychic structure of the human mind. According to Pestalozzi, Kerschensteiner, Dewey, and many others, effective education cannot rely only upon memory and logic – it has to involve the whole of the human mind. Pawłowski's geography with its exceptional attractiveness, its picturesque visions of landscape, was in the best position effectively to engage young people's emotions, imaginations, and feelings. In this way Pawłowski's geography was in step with the general principles of pedagogy and psychology. Regional geography, as a field concerning landscapes which are perceived and experienced in everyday life, could be an effective way of 'bringing the school nearer life'.

Pawłowski had realized that 'the mind of a child is quite different from the mind of a scientist', and he objected to the geographical orientation in education. Young people require first of all stimulation by the diversity of the world's phenomena, its mystery and beauty. Formal explanations should be introduced later. Priority, he felt, ought to be given to regional geography since it would be a very attractive field for young people. Geography as a field has to be the basis for designing a unified vision of the world. It could facilitate reconciliation of the scientific and humanistic visions of the world. Pawłowski's concepts with regard to political geography seem to be valuable from the educational point of view. At the dissolution of the Warsaw IGU Congress he said:

> 'In geography there exists a natural origin of internationalism and universality. Geography is the science of understanding and agreement between the

> nations. Repeatedly among geographers there have been so many apostles of universalism and the brotherhood of men.

Pawłowski believed that peaceful coexistence with others is the inclination of humanity. Geography, through its colourful depiction of foreign countries, could further develop such an inclination. He wrote, 'Geographical knowledge of foreign countries is the only serious basis for normalization of international relations.' Numerous of his textbooks were strongly influenced by such ideas. In 1938 he published *Geography as a Science and as a School Subject*. It contains a selection of papers published in the years 1913–38, concerning the problems of the essence of geography, its scope, aims and significance. Few are the geographers today who devote their effort and scientific creativity to the fundamental problems of geography, its vital concepts and methodological bases. Few are those who have given so much to education. In these areas no one equalled the contribution of Stanisław Pawłowski to Polish geography.

3. Influence and Spread of Ideas

Stanisław Pawłowski and his mentor, Eugeniusz Romer, were the leading Polish geographers of the interwar period. Pawłowski was a person, who in large degree, helped to create the image of Polish geography abroad. Thanks to his specialized and methodological works, he gained great recognition among geographers. His strong character and competence discouraged potential adversaries who might otherwise have wasted his time arguing about insignificant matters. Moreover, his uncommon kindness to, and spiritual community with, persons interested in geography brought him many friends and disciples. The concept of geography as a synthetic science based on elements of positivism as well as on some idealistic strains of philosophy, psychology and sociology found widespread acceptance in Polish society. The seed sowed by Pawłowski had fallen on good soil, and his views were in conformity with the intellectual atmosphere of the time. His *Geography of Poland*, edited in 1917 at roughly the time of independence, brought him popularity. After 1918 Polish social consciousness began to include once again the intellectual conceptions closely linked to nineteeth-century romanticism. Pawłowski's geography was an element of such a romantically influenced ideology. His school textbooks appeared in revised and updated versions, and were often reprinted. These books have been responsible for the education of a generation of Poles. The effectiveness of such an education was verified during the September Campaign of 1939, in the Warsaw Insurrection, and on all the battlefields of World War II.

Besides his broadly accepted methodological and pedagogical works, Pawłowski played an important role as an organizer and promoter of scientific investigations. Studies in rural settlement which he arranged in Poznań were developed successfully in other geographical centres, namely Kraków and Lwów. This culminated in some fifteen papers on the subject being presented during the Warsaw IGU Congress. Pawłowski was a master of a generation of geographers who accepted his formulated theoretical assumptions. His students became well-known professors of geography in postwar Poland, namely Rajmund Galon at the Nicholas Copernicus University, Toruń, Jan Dylik at the University of Łódź; and Maria Kiełczewska-Załeska at the Maria Skłodowska-Curie University, Lublin. It was to his credit that the new Adam Mickiewicz University in Poznań in a very short time became one of the most active centres of geographical thought in Poland. Pawłowski's concepts concerning the glacial morphology of the Carpathian Mountains and Greater Poland remain valid today. Many of his works were published abroad, namely in the *Bulletin de la Société Belge des Études Géographiques, Deutsche Rundschau für Geographie, Comptes Rendus du Congrès International de Géographie–Amsterdam*, and others. The very strong position which Pawłowski had gained in Polish and foreign geography was, perhaps, acknowledged by the Nazis. They could not see any other way of neutralizing Pawłowski's influence than to kill him. His wisdom and influence cost Pawłowski his life. That fate was shared by his son Przemysław and his daughter Wanda.

Both Nazi Germany and Soviet Russia were fearful of the intellectual and moral strength of the Polish people; consequently, Stanisław Pawłowski was not the only geographer whose life was tragically ended during the war. The remains of many other Polish geographers murdered in concentration camps and the Soviet Gulag still lie in unmarked graves. Polish geography lost many of its leading representatives. That is why the continuity of the Polish geographical tradition was disrupted. A second reason was the introduction of a totalitarian ideology, which restricted the free development of science and disturbed scientific methodology. Consequently, the name of Pawłowski was not often mentioned in postwar Polish geography, and his methodological concepts were largely ignored. The only areas in which Pawłowski's scientific output was recognized were geomorphology and settlement geography. The development of Polish postwar geography was characterized by strict specialization and none of Pawłowski's students who survived the war were in a position to propagate the methodological views of their master.

With sadness we must conclude that in contemporary Polish geographical circles geography in its classical understanding has been eliminated. Today geographers prefer specialist research, perceiving only single aspects of the geographical environment. They touch only separated points of reality and do not cover its entire surface. The privilege of controversy proper to the potential synthetic vision continually loses its place in competition with certainty, which is more nearly the product of specialized studies. This state of affairs is largely due to the difficulties in formulating the operational model of research unifying the natural and humanistic knowledge. In spite of that, many geographers continue to be fascinated by geography, and do not agree on the necessity of artificial limitations of interests to any single element of reality. Proof of this may be found in the animated discussion in current Polish geographical journals concerning the integration of geography and development of the regional concept. With this thrust the name of Pawłowski appears once more in the works concerning the methodology and pedagogy of geography. One of those who have helped change the direction and content of Polish geography is the last living disciple of Pawłowski, Professor Anne Dylikowa, of the University of Łódź. These changes are now possible given political and social changes in Polish society. Since our country has become a truly independent state and the notion of 'People's Republic' has been replaced by the 'Third Commonwealth', we can expect a return to the lost tradition of interwar geography and the restoration of Pawłowski's ideas.

Bibliography and Sources

1. REFERENCES ON STANISŁAW PAWŁOWSKI

Czekańska, M., 'Zasługi Stanisława Pawłowskiego na polu metodyki i dydaktyki geografii' (Pawłowski's merits in the field of didactic geography), *Czasopismo Geograficzne*, Vol. 17, 1936-46, pp. 3-11.

Dylik, J., 'Stanisław Pawłowski', *Rocznik Polskiego Towarzystwa, Geologicznego* (Annals of the Polish Geological Society), Vol. 19 (1950), pp. 215-30.

Młodziejowski, J., Dylik, J., Jurczyński, J., Romer, E., 'Stanisław Pawłowski, 1882-1939', *Przegląd Geograficzny*, Vol. 19 (1939-1945), pp. 52-77.

Olszewicz, B. (ed.), *Stanisław Pawłowski. W dwudziestą piątą rocznicę zgonu (1940-1965)* (On the 25th anniversary of the death), PWN, Warszawa, 1968, 188 pp. (contains full bibliography).

2. *SELECT BIBLIOGRAPHY OF WORKS BY STANISŁAW PAWŁOWSKI*

a. On Geomorphology and Hydrology

1910 'Die Temperatur fliessender Gewässer Galiziens', *Bul. Int. Ac. Sc. et Lettres*, Cracovie, Sér. A, pp. 625-32.

1911 'Zjawiska erozji na północnej krawędzi podolskiej' (Erosion on the northern Podole escarpment), *Kosmos*, Vol. 36, pp. 537-48.

1912 'Die Eisverhältnisse der oberen Weichsel, des oberen Dniestr und ihre Nebenflüsse', *Mitt. d. Geogr. Ges.*, Wien, Vol. 55, pp. 552-62.

1914 'Einiges über die Untersuchungsmethoden der Landformen', *Deutsche Rundschau für Geogr.*, Wien, Vol. 36, pp. 198-201.

—— 'Über ein altes Talstück in der Bukowina', *Mitt. d. Geol. Ges.*, Wien, pp. 246-55.

1915 'Ze studiów nad zlodowaceniem Czarnohory' (On the glaciation of Czarnohora region), *Prace Tow. Nauk. Warsz.*, (Studies of the Scientific Society of Warsaw), Vol. 10, 61 pp.

—— 'Z morfologii pienińskiego pasa skałek' (On the morphology of Pieniny rocky belt), *Kosmos*, Vol. 40, pp. 111-40.

1916 and J. Nowak, 'Tektonische und topographische Gliederung der Ostkarpathen', *Mitt. d. Geogr. Ges.*, Wien, Vol. 59, pp. 261-5.

1921 'O asymetrii dolin w dorzeczu Sanu' (On the asymmetry of valleys in the San basin), *Kosmos*, Vol. 46, pp. 492-502.

1922 'Charakterystyka morfologiczna wybrzeża polskiego' (Morphology of the Polish seashore), *Prace Pozn. Tow. Przyj. Nauk.*, Works of Poznanian Friends of Science Association, Ser. A, Vol. 1, pp. 1-90.

—— 'O utworach na dnie zatoki Gdańskiej' (On the sedimentation of Gdańsk Bay), *Prace Pozn. Tow. Przyj. Nauk.*, Ser. A, Vol. 1, pp. 143-166.

1923 'O pojmowaniu terasy rzecznej w morfologii' (On the understanding of the 'river terrace' notion in morphology), *Kosmos*, Vol. 48, pp. 576-83.

1925 'Sur les traces glaciaires dans les Gorganes occidentales (Karpates polonaises)', *Bul. Int. Ac. Pol. Sc. et Lettres*, Ser. A, pp. 363-75.

1928 'O podziale polskich Karpat Wschodnich' (On the division of the Polish

Eastern Carpathian Mts), *Czasopismo Geograficzne*, Vol. 6, pp. 77–86.

—— 'Pleistocene terraces in Poland', Intern. Geogr. Congress, *Reports*, Cambridge, pp. 167–8.

1929 'Rozważania nad morfologią doliny Warty pod Poznaniem' (Considerations concerning the morphology of Warta valley near Poznań), *Bad. Geogr. nad Polską Płn.-Zach.* (Geographical investigations in north-western Poland), Poznań, Vols 4–5, pp. 91–106.

1930 'Les terrasses pléistocènes en Pologne', in *2 Rapport Comm. Terr. Plioc. Pléist.*, Florence, pp. 108–12.

1931 'O kształtach powierzchni i o podziale Wielkopolski na krainy' (On the morphology and regional division of greater Poland), *Bad. Geogr. nad Polską, Płn.-Zach.* (Geographical investigations in north-western Poland), Vol. 6–7, pp. 137–72.

1934 'Les Karpates à l'époque glaciaire', *C. r. Congr. Int. Géogr.*, Varsovie, Vol. 2, pp. 89–149.

1936 'Lodowiec Blaaisen w Norwegii' (The Blaaisen Glacier in Norway), *Czasopismo Geograficzne*, Vol. 15, pp. 123–35.

1938 'Le problème des moraines terminales', *C. r. Congr. Int. Géogr.*, Amsterdam, Vol. 2, pp. 230–50.

b. On Human Geography and Regional Geography

1911 'Niektóre kanały spławne na ziemiach polskieh' (Navigable channels on Polish territory), *Spraw. Ak. Handl.*, Reports of the Academy of Commerce, Lwów, 32 pp.

1913 'Geografia fizyczna ziem polskich według Encyklopedii Polskiej' (Physical geography of Polish territory according to the *Polish Encyclopaedia*), *Kosmos*, Vol. 38, pp. 488–509.

1917 *Geografia Polski* (Geography of Poland), Książnica Polska, Lwów, 203 pp.

1920 *Territoire plébiscitaire sur la rive droite de la Vistule*, Poznań, 22 pp.

1924 'Położenie geograficzne Polski' (Geographical situation of Poland), *Czasopismo Geograficzne*, Vol. 2, pp. 435–41.

—— 'Znaczenie Bałtyku' (The significance of the Baltic Sea), *Ziemia*, Vol. 9, pp. 62–9.

—— 'Zmiany w ukształtowaniu powierzchni ziemi wywołane przez człowieka' (Changes in morphology caused by human activity), *Przeglad Geograficzny*, Vol. 4, pp. 48–64.

1925 'Geograficzny krajobraz północnej części oazy Kharga' (Geographical landscape of the northern part of the Kharga oasis), *Czasopismo Geograficzne*, Vol. 3, pp. 18–38.

1926 'O rozmieszczeniu ludności polskiej w woj. pomorskim' (On the distribution of the Polish population in Pomerania), *Bad. Geogr. nad Polską Płn.-Zach.* (Geographical investigations in north-western Poland), Vols 2–3, pp. 89–103.

1927 'Geograficzny krajobraz terytorium W.M. Gdańska' (Geographical landscape of the Gdańsk Independent City), *Rocznik Gdański*, Vol. 1, pp. 3–21.

1929 'Walia jako indywidualność geograficzna' (Wales as a geographical individuality), *Przeglad Geograficzny*, Vol. 9, pp. 197–226.

1934 'L'habitat rural en Pologne', *C. r. Congr. Int. Géogr.*, Varsovie, Vol. 3, pp. 497–507.

1936 'Geografia polityczna Polski' (Political geography of Poland), in S. Pawłowski, J. Bystroń, A. Peretiatkowicz (eds), *Polska współczesna* (Contemporary Poland), fifth edition, Lwów.

c. On Methodology and Teaching

1914 'W sprawie polskiej nomenklatury geograficznej' (On Polish geographical terminology), *Kosmos*, Vol. 39, pp. 615–55.

1918 'Antropogeografia' (Human geography), *Przeglad Geograficzny*, Vol. 1, pp. 121–8 and 311–21.

1919 'O potrzebach geografii polskiej' (On the needs of Polish geography), *Nauka polska*, Vol. 2, pp. 64–9.

1932 'Tendances nouvelles en géographie', *Bul. de la Soc. Belge d'Études Géogr.*, pp. 1–8.

1938 'Inwieweit kann in der Anthropogeographie von einer Landschaft die Rede sein?', *C. r. Congr. Int. Géogr.*, Amsterdam, Vol. 2, pp. 202–8.

—— 'O fikcjach w nauczaniu geografii' (Fictions in geographical teaching), *Czasopismo Geograficzne*, Vol. 16, pp. 1–10.

—— *Geografia jako nauka i przedmiot nauczania* (Geography as a science and as a school subject), Książnica-Atlas, Lwów, 215 pp.

3. ARCHIVAL SOURCES

The correspondence of Stanisław Pawłowski is to be found in the state archives of the City and Voivodship of Poznań. There are deposited about 3000 letters to Pawłowski as well as copies of his replies.

Witold Wilczynski is a lecturer at the Jagiellonian University in Kraków, Poland.

Chronology

1882 Born in Dembowiec, Galicia, 16 March

1902–08 Studied at the universities in Vienna and in Lwów

1910 Completed Ph.D. degree at the John Casimir University in Lwów

1917	Published *Geography of Poland*
1918	Professor of geography at the University of Lwów
1919	Moved to Poznań and became a director of the Geographical Institute at Adam Mickiewicz University
1922	Founder and chairman of the Association of Polish Geography Teachers
1924	Extended fieldwork in Egypt
1928	Fieldwork in Britain
1930	Arranged the Faculty of Didactics at the Geographical Institute, University of Poznań
1932–33	Rector of Adam Mickiewicz University in Poznań
1934	Secretary General of the Organizing Committee of the Fourteenth International Geographical Congress in Warsaw
1938	Elected vice-president of the International Geographical Union; publication of *Geography as a Science and as a School Subject*
1940	Murdered in Seventh Fort of Poznań, January

Raphael Pumpelly

1837–1923

Margaret D. Champlin

Although Raphael Pumpelly was trained as a mining geologist, he had a varied career that included activities relating to both physical and human geography. He was a state geologist, a division head with the US Geological Survey, and an explorer and organizer of scientific expeditions that provided research opportunities to other scientists. His explorations in Turkestan in 1903 and 1904 were pioneering efforts to correlate the geomorphic evolution of central Asia with the factors that brought about the beginnings of civilization.

1. Education, Life and Work

Pumpelly was born in Owego, New York, on 8 September 1837, the youngest son of a transplanted New Englander whose family pioneered lumbering, surveying and land development in the southern part of the state. He was an adventurous child and spent much time out of doors exploring the river, hills and ravines near his home, but he also received the loving attention of his mother, a cultured woman who was dissatisfied with her small-town life. After preparation for Yale University at a New Haven, Connecticut, school, Pumpelly persuaded his mother to take him to Europe to study instead of sending him to Yale. He attended technical schools in Hanover and Paris, tramped in the Alps, and independently explored Vesuvius and the island of Corsica, acquiring in a rather haphazard way quite an extensive knowledge of geology. Various scientists who recognized his talents encouraged his geological studies, and, as a result, Pumpelly decided, in 1866, to enroll at the Royal School of Mines at Freiberg, Saxony, which was to attract many American students over the next few decades.

Pumpelly's training at one of the best mining schools of his day not only prepared him for a career as a mining geologist, but it also encouraged his interest in a broader range of subjects that continued throughout his life. His professor of geology, Karl Bernhard von Cotta, was a pioneer in the study of petrology and the science of ore deposits, and was also recognized as a superb map maker and a student of human geography. In lectures and on field trips, Pumpelly must have absorbed Cotta's belief that geological change had profoundly affected the development of all living things. Before leaving Freiberg in 1859, Pumpelly published his first scientific paper, based on his observations of glacial activity on Corsica (Pumpelly, 1858).

In 1860, Pumpelly accepted a position as a mining engineer at an Arizona silver mine, in territory recently acquired from Mexico through the Gadsden Purchase. His success in adapting Freiberg methods under the difficult conditions imposed by the Sonoran Desert and his courage in resisting Apache attacks impressed Josiah Dwight Whitney of the newly organized California Geological Survey. In 1861, Whitney recommended Pumpelly for a position as mining geologist and engineer with the Japanese Shogun government, which was embarking on a programme to improve its mining efficiency by bringing in Western experts. Because of political unrest, Pumpelly and his partner, the geologist William Phipps Blake, were limited to working on the mostly undeveloped island of Hokkaido (then Yesso) and teaching a few students who acted as their assistants. They explored southern Yesso, observing and assessing gold, lead, sulphur, and silver mines, their principal contribution being to recommend the use of pumping machinery and to introduce blasting.

Increasing opposition to foreigners brought Pumpelly's work for the Japanese government to an abrupt end in 1863, but his sense of adventure led him on a journey of geological and geographical exploration that took him to the interior of China and across Mongolia before he returned to America in 1865. He travelled up the Yangtze and explored along the Great Wall and the edge of the Mongolian plateau, noting the enormous loess deposits there and attempting to develop a theory of their origin. He examined coal-mines north-west of Peking for the Chinese government, and his report, estimating the extent of China's rich coal resources and making recommendations for their more efficient exploitation, was sent to US Secretary of State Seward and appeared in mining publications in Britain and France.

Returning to America in 1865, Pumpelly was invited by Whitney to present his findings on China and Japan to the newly founded National Academy of Sciences (Pumpelly, 1866). Because so little was known of the geology of Japan and China and because of the great interest in their mineral resources, Pumpelly prepared a much expanded version of the Academy paper for Joseph Henry of the Smithsonian Institution (Pumpelly, 1867). This report included the first coloured geological map of China. Pumpelly also wrote a popular version of his travels, which established his reputation as an adventurous scientific explorer. *Across America and Asia* (Pumpelly, 1870) contained valuable observations on the geography and customs of the people of Japan and China as well as a persuasive chapter on the desirability of allowing Chinese immigration to the US.

Pumpelly's professional work for the next few years was in the field of economic geology. He investigated a geological problem encountered by builders of the Hoosac railway tunnel in western Massachusetts and explored for minerals in northern Michigan for a group of Boston investors. From 1871 to 1872 he worked for the Michigan Geological Survey in the copper region of the Upper Peninsula, doing pioneering studies on the origin of copper ores and the paragenesis of copper minerals (Pumpelly, 1871 and 1873). From 1872 to 1873 he headed the Missouri Geological Survey, where he learned how unreliable state legislatures could be as patrons of science in the days before adequate funding was built into the budget (Pumpelly, 1873a).

Retiring from the Missouri Survey for health reasons, Pumpelly spent the mid-1870s pursuing private research. He took up scientific farming in New York state and devoted himself to the study of petrology, adopting the then new methods of microscopic petrography. With the use of the polarizing microscope, he examined hundreds of specimens of rock in thin section, which he prepared himself, in order to study the changes which minerals had undergone (Pumpelly, 1878). It is a pioneering study on the origin of ores that indicates his enthusiasm for exploring new intellectual fields as well as geographical ones. It was at this time that he also did the reading and research that led to his adopting and modifying Ferdinand von Richthofen's theory of the origin of loess (Pumpelly, 1879).

In 1879, Clarence King became director of the new US Geological Survey (USGS), an organization then primarily orientated toward serving the rapidly growing mining industry, and King chose Pumpelly to head the division of mining geology. One of the Survey's first undertakings was to collect data for the US Tenth Census, which was greatly expanding its coverage under Census Superintendent Francis Amasa Walker. For two years Pumpelly was in charge of compiling statistics on the mineral industries of the USA, exclusive of the precious metals. He directed and co-ordinated a team of experts who often personally visited remote mining establishments with their official questionnaires. Although his Tenth Census *Report* (Pumpelly, 1886) contained data on all minerals, Pumpelly placed special emphasis on the iron ores, which he felt were vital to the country's development and whose extent was not well understood.

The Tenth Census work was not yet finished when, in 1881, Pumpelly resigned from the USGS to direct a geological and geographical survey of the Northwest for the financier Henry Villard, who had recently gained control of the Northern Pacific Railroad. Villard wanted a thorough scientific exploration of the railway's land grant from the state of Washington east to the Dakotas that would guide the railway in building feeder lines.

Pumpelly hoped that his survey would promote the rational development of the Northwest. In some respects his goals resembled those of John Wesley Powell, who, in his *Report on the Lands of the Arid Region* in 1878, had presented to Congress a radical new proposal for the development of western lands. Pumpelly's plan was not only to survey timber and mineral lands, but also to investigate the agricultural potential of the semi-arid region, including determination of lands suitable for irrigation agriculture and grazing.

The Northern Transcontinental Survey (NTS) team was composed of experts in many fields, organized into seven divisions covering mineral resources, climate and rivers, soils, forests, economic botany, chemistry and topography. The division of topography, under Allen D. Wilson, had the task of preparing accurate topographical maps of the Northwest, on which were to be recorded the data obtained by the other divisions. The division of climate and rivers, headed by Edward Singleton Holden, compiled all previously recorded meteorological data for the Northwest and did preliminary work on the development of methods for stream measurement. The division of soils was directed by W. Eugene Hilgard, of the University of California, who mapped the Yakima and Colville regions, indicating the areas that were irrigable, partially irrigable, suitable for grazing, or unsuitable for either irrigation or grazing (Pumpelly, 1882–83). The division of forests was headed by Charles Sprague Sargent, director of Harvard's Arnold Arboretum and the country's foremost authority on forestry. He saw his work with Pumpelly as a means of gathering data on western trees for his monumental *Silva*. Forage plants were to be studied by

W.M. Canby, in charge of the division of economic botany; and insects injurious to vegetation and timber were to be investigated by H.A. Hagen.

Pumpelly made three explorations of the territory for purposes of reconnaissance during the summers of 1881, 1882 and 1883. In 1882 and 1883, his party entered what is now Glacier National Park, and in 1883 he made the first sighting of a glacier in the Rockies, for which a glacier in the park has been given his name. Two members of the NTS, Charles Sargent and Bailey Willis, were active in promoting government acquisition of the land, which became a national park in 1910.

Pumpelly's enlightened ideas for development of the Northwest were supported by the railway as long as Villard remained president, but in December 1893, on the verge of bankruptcy, Villard was forced to resign. Pumpelly tried to persuade the new directors to continue the survey but they failed to appreciate what he was trying to accomplish. The results of his survey were never written up in full. Some of the findings appeared as part of his Tenth Census volume, several of his men reported on their investigations in various scientific journals, and a summary of the NTS work appeared in George Merrill's *History of American State Geological and Natural History Surveys* (Pumpelly, 1920).

In 1884, Pumpelly returned to the US Geological Survey, which was growing stronger under the direction of John Wesley Powell, and stayed with it until 1892, when mandated cutbacks forced a drastic reduction of personnel. Powell placed greater emphasis on topographical mapping, general geology, and palaeontology than on economic geology. He put Pumpelly in charge of the division of Archaean geology and assigned him to investigate the older rocks of the Green Mountains in western Massachusetts. This region was at the centre of what had long been known as the 'taconic question', one of the most controversial problems of American geology. Pumpelly and his assistants spent the next several years working out the structure of the Green Mountains, producing a report that is still the classic starting-point for any structural study of that region (Pumpelly, 1894).

In 1902, Pumpelly began his pioneering geoarchaeological expeditions to Turkestan, supported by a grant from the newly established Carnegie Institution of Washington, DC. He was inspired by a long-standing interest in a question that puzzled so many of his contemporaries: the geographical origin of the Aryans, from whom, it was believed, the Indo-European languages had sprung. He agreed with one popular hypothesis which held that the Aryans had come from Central Asia, and he thought he could shed light on the question by means of a well-planned and thorough investigation of past climatic conditions in the Trans-Caspian region, combined with archaeological excavations at promising sites.

Although Pumpelly's explorations in Turkestan in 1903 and 1904 are remembered primarily for the discovery and excavation of a prehistoric culture at Anau, near Ashkhabad, they were the work of an interdisciplinary team of specialists unusual for that time. His final reports included papers by physical geographers, an anthropologist, a botanist, a comparative osteologist, and a metallurgist (Pumpelly, 1905 and 1908). William Morris Davis was a member of his party in 1903 and Ellsworth Huntington in 1903 and 1904. The archaeological work at Anau, under the direction of Hubert Schmidt, was a model of scientific excavation, and for many years Anau was the representative site for a primitive Central Asian culture.

Pumpelly's research in the physical geography and geology of the region, carried out in parallel with the archaeological excavations, established to his satisfaction that there had been a close relationship between glacial activity, the amount of erosion in the mountains, the rate of growth of alluvial fans, and periods of growth or decline of cultures. The archaeological work indicated that the earliest mound dwellers at Anau had practised a primitive form of agriculture, had learned to domesticate certain breeds of animals, and had worked in copper. 'In Anau I,' he asserted, 'we have the oldest instance of an organized agricultural civilization at present known' (Pumpelly, 1918, Vol. II, 808–9).

A third expedition was planned for 1905, but political events in Russia prevented Pumpelly's return to Central Asia. Until 1908 he was happily engaged in the study and research necessary for the preparation of his final report on the expedition of 1904. His wife died in 1915, and some time after that he began the writing of his memoirs for his grandchildren, which developed into the well-reviewed, two-volume *My Reminiscences*, based in part on his earlier *Across America and Asia*. He died on 10 August 1923, at the age of eighty-five, and is buried with his wife in Newport, Rhode Island.

2. Scientific Ideas and Geographical Thought

Pumpelly believed that climatic and geological change had affected the development of civilization; that primitive human societies had responded to changing conditions with new inventiveness and had spread their cultural accomplishments through migration. In *Across America and Asia* and in his *Reminiscences*, Pumpelly briefly discussed how geography had influenced the prehistoric Indian populations of the Southwest or the development of Chinese civilization, but he expressed the concept most fully in the report of his 1904 expedition to Turkestan (Pumpelly, 1908).

Pumpelly's geographical and geoarchaeological investigations in Central Asia, inspired by his desire to trace the origins of the Indo-European languages, became an ambitious regional study involving the interrelation of physiography and the evolution of cultures. He thought of the whole inner continent as an organism in which all geographical processes were interrelated. The glaciers of the last ice age had fed an inland sea larger than the Mediterranean, which he believed had supported a thriving primitive society. As the glacial period passed and evaporation became more rapid than the inflow of water, the rivers and the sea had shrunk and the dried silts were dispersed by the winds. The finest materials were deposited as loess, while the coarser sands formed dunes that encroached on previously fertile lands. With the progressive aridity, life became restricted to the mountains and the bordering alluvial fans, where water was still available (Pumpelly, 1908, pp. 11–13). It was on one of these fans, or 'delta oases', as he called them, that he conducted his geoarchaeological excavations.

The whole innercontinental region was 'predestined to a definite course of life history', he thought. Although this could be interpreted as a rigid environmental determinism, his thinking is perhaps closer to William Morris Davis's concept of ontography or Arnold Toynbee's theory of 'challenge and response'. Toynbee, in fact, later used the results of Pumpelly's investigations to show that the nomads of Central Asia had developed their way of life in response to the challenge of the arid environment. For Pumpelly, the geographical features became the 'motivating factors in the evolution of the intellectual and social life of man'. The early oasis dwellers responded to the challenge of the arid environment by developing the skills that allowed them to survive, and transmitted these skills to Europe through waves of aridity-propelled migration. His migration theory remained for him a hypothesis only, since he felt it was unsubstantiated by his geoarchaeological research, and he withheld from publication the chapter on Aryan migrations he had written for his Carnegie report.

At the heart of Pumpelly's version of the evolution of the arid regions of central Asia were his ideas about the origin of loess. He had become interested in the question of its origin when he was in northern China in 1863. He had then thought that the 'terrace loam', as he called it, which he saw mantling the sides of mountains and filling ravines to a depth of several hundred feet, had been deposited in a great freshwater lake. The lacustrine theory was also held by American geologists who were familiar with the 'bluff deposits' of the Mississippi Valley, which Lyell had earlier equated with the loess deposits along the Rhine.

In 1878, Pumpelly became converted to the aeolian hypothesis of Ferdinand von Richthofen that is generally accepted today when he reviewed the first volume of Richthofen's *China* for *The Nation* (Pumpelly, 1878a). He was convinced of the truth of the theory by Richthofen's discovery of the shells of land snails in the loess at elevations of 18 000 feet, and by the 'fearful' dust-storms he had himself witnessed in China.

In an article written the following year (Pumpelly, 1879), Pumpelly supplemented Richthofen's theory with a new explanation of the original source of the fine material of which loess is composed. He argued that glacial grinding, which Richthofen and others had proposed, could not have produced enough fine material to account for the enormous extent of the loess in North America and China. He suggested a second source, based on personal observation and his extensive knowledge of petrography. He believed that much of the material must have come from a long process of rock decay caused by the carbonic acid generated by overlying vegetation. This process would have decomposed granites and gneisses and left masses of fine insoluble clays and sands surrounding the undisturbed rock cores, protected from erosion by abundant vegetation. Pumpelly had observed such decomposed masses in railway cuts in unglaciated areas of the eastern US states. In the northern part of the country the decomposition products had been removed by glaciers, but in regions where the climate had become dryer and there was less protective vegetation, the fine materials had become the prey of the winds.

Pumpelly's approach to geographical questions was almost invariably that of the geologist. In his explorations in Arizona, Japan, China, the Pacific Northwest, or Turkestan, his field notebook was always at hand for recording the nature of the geological formations he encountered. His major study of the structure of the Green Mountains for the US Geological Survey was based on the most meticulous of geological and petrographical investigations. He thought that only the kind of detailed study of ancient shorelines that Grove Karl Gilbert had done at Lake Bonneville in the 1880s could determine periods of expansion or shrinking of the Caspian. He studied the past physical geography of Anau by digging shafts in the vicinity of the mounds and analysing the sediments in order to determine the relation of cycles of wet and dry periods to periods of build-up or erosion of the delta oasis. He even saw the archaeological excavations at Anau as a geological problem, since he felt that the artefacts uncovered by archaeologists may be used to determine time sequences in the same way that fossils are used by geologists.

And yet his geological perspective did not obscure his broader perceptions. He had an exceptional ability to visualize all of the processes that had been at work in the evolution of the total landscape. Whether he was studying the structure of the Green Mountains, surveying the economic resources of the Northwest, or investigating the culture and environment of postglacial man in central Asia, he brought together the interrelated aspects of his subject with an outlook characteristic of a geographer.

3. Influence and Spread of Ideas

Pumpelly first made a name for himself as an explorer, and he continued to exert his influence principally through his organization and administration of scientific surveys and expeditions. His explorations in China and Mongolia in the early 1860s, before any other American geologist had been there, and his popular and scientific writings about his travels, no doubt contributed to the desire of Americans to accumulate scientific data about little-known foreign countries as well as about the unexplored parts of the American West.

Since Pumpelly did not have an extensive teaching career, much of his influence was felt in day-to-day contact in the field with members of his survey teams and research expeditions, in the form of suggestions on the way a problem might be approached. Several of his assistants acknowledged that they owed much to his insight into geological processes. Bailey Willis, who had worked with Pumpelly on a number of surveys and looked on him as a mentor, felt that it was 'in conversation in the field face to face with the facts or in talks of an evening between days' work' that Pumpelly made some of his greatest contributions (Willis, 'Memorial', 75–6).

Most people found it stimulating to work on one of Pumpelly's survey teams, which resembled in some respects a research school in a university. He was a strong and charismatic research director, who drew up the plans for attacking a problem and chose people who shared his research goals but also encouraged independent work and publication. One postgraduate student who worked with him for a summer in the Green Mountains remembered that it was an experience worth several semesters in a university.

Both William Morris Davis and Ellsworth Huntington found that being with Pumpelly in Central Asia stimulated their interest in human geography and allowed them opportunities to pursue their own research. Davis's participation in the Trans-Caspian expedition, for example, gave him a chance to study cycles of erosion in a glacial environment and to collect examples of ontography, at a time when he was just beginning to develop his ideas on the subject. Huntington's research in Turkestan was the beginning stage in the development of his ideas on climatic pulsations. Pumpelly gave Huntington the chance to spend two years in the Asia he loved. During this period of close association there must have been many occasions when Pumpelly's suggestions and counsel would have provided Huntington with new insights into the problems of human geography that were to occupy him for the rest of his life.

Bibliography and Sources

1. SELECTED BIBLIOGRAPHY OF WORKS ABOUT RAPHAEL PUMPELLY

Amstutz, G.C., 'Raphael Pumpelly', *Dictionary of Scientific Biography*, Vol. 11, 209–11, New York: Scribner's (1975).

Champlin, Margaret D. 'Raphael Pumpelly and American Geology in the Gilded Age', Ph.D. diss., UCLA (1989).

Keyes, Charles, 'Raphael Pumpelly: Premier Explorer', *Pan-Am. Geologist*, Vol. 40 (1923), 241–50.

Merrill, George P., 'Raphael Pumpelly', *Dictionary of American Biography*, Vol. 8, 264–6, New York: Scribner's (1935).

Pumpelly, Raphael, *My Reminiscences*, 2 vols, New York: Henry Holt (1918).

Willis, Bailey, 'Memorial of Raphael Pumpelly', *Bull. Geol. Soc. Am.*, Vol. 36 (1925), 44–84.

Pumpelly's papers are in the Henry E. Huntington Library, San Marino, CA. They include field notebooks, a limited amount of personal correspondence, and many letters of family members. Other Pumpelly papers at the Huntington are in the Willis (Bailey) Collection, Section III, Box 28. They include correspondence and records collected by Willis for his memorial of Pumpelly. With the Willis papers is Pumpelly's unpublished chapter entitled 'The Migrations and the Aryan People', originally intended for inclusion in Explorations in Turkestan; Expedition of 1904. (The chapter is reproduced in Champlin, Appendix II, 425–60.)

2. SELECTED BIBLIOGRAPHY OF WORKS BY RAPHAEL PUMPELLY

1858 'Über einige Gletscher-Überreste der Insel Corsica', *Neues Jahrbuch für Mineralogie, Geognosie, Geologie*, 273–7.

1866 'Notice of an account of geological observations in China, Japan and Mongolia', *Am. J. Sci.*, ser. 2, Vol. 41, 145–9.

1867 'Geological researches in China, Mongolia and Japan, during the years 1862 to 1865', *Smithsonian Contributions to Knowledge*, Vol. 15, Washington, DC: Smithsonian Institution.

1870 *Across America and Asia; Notes of a Five Years' Journey Around the World, and of Residence in Arizona, Japan and China*, New York: Leypoldt and Holt.

1871 'Paragenesis and derivation of copper and its associates on Lake Superior', *Am. J. Sci.*, ser. 3, Vol. 2, 188–98, 243–58, 347–55.

1873 'Copper District', in Geological Survey of Michigan, *Upper Peninsula, 1869–1873*, Vol. 1, part II, New York: Julius Bein.

1873a Geological Survey of Missouri, *Preliminary Report of the Iron Ores and Coal Fields from the Field Work of 1872*, New York: Julius Bien.

1878 'Metasomatic development of the copper-bearing rocks of Lake Superior', *Am. Acad. Arts Sci. Proc.*, Vol. 13, 253–309.

1878a 'Richthofen's China', *The Nation*, Vol. 26, 231–2, 243–4.

1879 'The relation of secular rock-disintegration to loess, glacial drift and rock basins', *Am. J. Sci.*, ser. 3, Vol. 17, 133–44.

1882–83 Northern Transcontinental Survey, *Map Bulletins*, New York: Julius Bien.

1886 Bureau of the Census, *Report on the Mining Industries of the United States (Exclusive of the Precious Metals)*, US Tenth Census, Vol. 15, Washington, DC, Government Printing Office.

1894 Raphael Pumpelly, J.E. Wolff and T. Nelson Dale, *Geology of the Green Mountains in Massachusetts*. (USGS Monograph No. 23), Washington, DC: Government Printing Office.

1905 Raphael Pumpelly (ed.), *Explorations in Turkestan, with an Account of the*

Basin of Eastern Persia and Sistan. Expedition of 1903 (CIW Publication No. 26), Washington, DC: Carnegie Institution of Washington.

1908 Raphael Pumpelly (ed.), *Explorations in Turkestan; Expedition of 1904. Prehistoric Civilizations of Anau; Origins, Growth and Influence of Environment*, 2 vols (CIW Publication No. 73), Washington, DC: Carnegie Institution of Washington.

1918 *My Reminiscences*, 2 vols, New York: Henry Holt.

1920 'The northern transcontinental survey', in George Merrill, *Contributions to a History of American State Geological and Natural History Surveys* (US National Museum Bulletin 109), Appendix 2, pp. 539–44, Washington, DC: Government Printing Office.

Margaret Champlin was science librarian at California State University, Los Angeles, from 1972 to 1986. She received her Ph.D. in history from UCLA in 1989, on completion of her dissertation on Pumpelly. She is an independent scholar living at 2169 Linda Flora Drive, Los Angeles, CA 90077.

Chronology

1837	Born on 8 September in Owego, New York
1854–56	Travelled with mother in Germany, France and Italy, studying intermittently at various technical schools
1856–59	Attended Royal Academy of Mines at Freiberg, Germany
1860–61	Worked at Santa Rita silver mine near Tubac, Arizona
1861–63	Served as mining expert to Japanese government
1863–65	Explored in China and Mongolia
1867	Publication of *Geological Researches in China, Mongolia and Japan* by Smithsonian Institution
1869	Married Eliza Shepard of Dorchester, Massachusetts
1870	Publication of *Across America and Asia*
1870–71	Surveyed copper region of Northern Peninsula for Michigan Geological Survey
1872	Elected member of the National Academy of Sciences
1872–73	Directed Missouri Geological Survey
1873–79	Lived as gentleman farmer, doing private research and consulting
1879	Joined US Geological Survey (USGS)
1880–81	Directed survey of mineral resources for US Tenth Census
1881–84	Directed Northern Transcontinental Survey for Northern Pacific Railroad
1884–92	With USGS; directed structural study of Green Mountains

1902–04	Led Trans-Caspian Expeditions to Central Asia for Carnegie Institution; conducted geoarchaeological studies at Anau
1918	Publication of *My Reminiscences* by Henry Holt
1923	Died 10 August in Newport, Rhode Island

Josef Schmithüsen

1909–1984

Emil Meynen

1. Education, Life and Work

Josef Schmithüsen was born on 30 January 1909 in Aachen, the oldest son of a merchant. Schmithüsen attended secondary school in Aachen, taking classes with two famous teachers, the biologists Franz Roth and Mathias Schwickerath, the latter later appointed Professor of Plant Geography at the University of Bonn. Schmithüsen's teachers gave him the knowledge in the fields of vegetation ecology and forestry which influenced his choice of these aspects in his studies. Schmithüsen studied for three semesters at the Technical University in Aachen (1927–29), in the fields of chemistry, physics and geology. His main interest was the field of botany; in 1929 he was the youngest member of the Botanical Excursion of the State Conservation Office. During the excursion, Schmithüsen became acquainted with famous biologists, viz. Reinhold Tuxen, Holzenau (Weser), K. Hueck (Halle), and J. Braun-Blanquet (Montpellier). These scientists encouraged him to carry out his university studies at Bonn. At Bonn University, Schmithüsen attended Alfred Philippson's last lecture, 'Climate and Vegetation of the Earth'. Schmithüsen was impressed by the interaction of climate, vegetation and animal life as explained in the lecture, aspects which he had never before considered. He determined to become a geographer. His teacher was Leo Waibel, Philippson's successor, who taught Schmithüsen to observe natural landscapes and develop the talent of observation of natural regions.

Schmithüsen took his doctoral degree in 1932, the subject of his thesis being 'The Woodlands of the Western Rheinish Uplands'. The following year, he was granted a scholarship to carry out the first geographical vegetation survey through southern France to eastern Spain. Schmithüsen carried out fieldwork for almost seven years in the Grand Duchy of Luxembourg, and completed his inaugural dissertation in 1939 on a regional geographical theme: 'Luxembourg – Land, Natural Geography, National Groups and Rural Economy'.

Schmithüsen's interest in the theme was mainly devoted to a physiogeographical classification on the basis of vegetation, historical-cultural elements, and rural landscapes. Schmithüsen taught geography at Bonn University for nine semesters; in 1941 he took up the position of co-worker and assistant director at the newly founded Department of Regional Geography, National Office of Cartography, in Berlin, under the direction of Emil Meynen. Schmithüsen was the initiator of the concept of 'Natural Regions', which was applied to Meynen's aim of developing a geographical mapping system.

The work was interrupted by Schmithüsen's being conscripted and sent to the Eastern Front. Through the influence of Reinhold Tüxen, Schmithüsen was transferred to the Army's research branch in 1943, and worked as scientific liaison officer for the various working-groups preparing 'terrain maps'. The use of aerial photography for terrain mapping added another area of expertise to Schmithüsen's knowledge. Schmithüsen was a prisoner of war for two years, and lost all his belongings and property; after his release in 1947, he went to Karlsruhe and took up a lectureship at the university. Here, Schmithüsen rebuilt the Geographical Institute, which had been destroyed during the war. In 1953 Schmithüsen was appointed associate professor; in 1959 he was given tenure as professor of geography at the University of Karlsruhe.

Schmithüsen again took up the work of regional geographical mapping for the map system on the scale 1: 200 000, carrying out the topographical survey for the map sheet of Karlsruhe. Schmithüsen worked as co-author on the *Handbook of Natural Regions of Germany* (1953), which had been initiated by Emil Meynen; Schmithüsen was the author of the historical and methodological introduction. The original concept of natural regions had been changed by the application of landscape ecological methods, which Schmithüsen supported as a positive extension of the basic system. The success of the system of natural regions is evidenced by its use in the field of statistics and planning of historical and areal studies.

Ruthard Ohm, a geographer who had studied under Alfred Hettner, came to Karlsruhe in 1952 and took over Schmithüsen's position so that Schmithüsen was able, with the support of a grant provided by the German Research Society (Deutsche Forschungsgemeinschaft), to carry out a year-long vegetation geography study tour in Chile. The results of Schmithüsen's researches were presented in a series of articles on the forest communities of Chile. The articles treated mainly the evergreen sclerophyllous woods of central Chile; also a general geographical vegetation classification system was set up on the basis of natural regions for Chile.

It appeared that Schmithüsen would continue to devote his professional life to Karlsruhe; however, during a visiting professorship at the University of Auckland, New Zealand, Schmithüsen was offered a position as the second chair of geography at the University of Saarbrücken, working together with Carl Rathjens. Schmithüsen worked actively and productively in his position as professor of geography until his retirement in 1977. In his lectures, Schmithüsen presented both material from expert publications as well as the results of his own practical researches, illustrating the areas where geographical knowledge was still very sparse. He presented practical courses and carried out many field excursions throughout Europe, which his students found extremely valuable for their studies.

With the expansion of Saarbrücken University, Schmithüsen founded the Biogeographical Institute, whose director was P. Müller, and the journal *Biogeographica*. Up to 1970, Schmithüsen was chairman of the Institute of Regional Geography of Saarland. He was also responsible for the creation of a professorship of aerial photographic research. In alternate years Schmithüsen and Rathjens were institute directors at the university. This gave Schmithüsen the opportunity to carry out his numerous research expeditions, the results of which not only improved his own knowledge about the areas visited, but were also presented in his university courses. New knowledge regarding vegetation ecosystems prompted Schmithüsen to extend his scientific expeditions to the Pacific coastal lands - New Zealand, eastern Australia, Japan and California. Schmithüsen also continued his expeditions to Chile, where his detailed researches were devoted to the Chilean ecosystems.

Schmithüsen held a lectureship at the University of Toronto, Canada, in 1973; in 1976 he took part in an expedition through Central Asia - Tashkent, Samarkand and Bukhara - after attending the Twenty-third International Geographical Congress in Moscow. Participation at national and international symposia, and taking part in conferences at the Academy in Pressburg (Bratislava), Czechoslovakia, gave Schmithüsen the opportunity to discuss scientific aspects with colleagues from other countries. In 1970, at the symposium 'Biogeography and Ecological Landscape Research', sponsored jointly by the Max Planck Institute of Limnology (Section: Ecology of the Tropics) and the Geographical Institute of Saarbrücken University, Schmithüsen gave a special direction to the general theme with his paper 'Biogeographical Aspects and Methods in Geosystem Typology'.

2. Scientific Ideas and Geographical Thought

Schmithüsen readily agreed to write the volume *Plant Geography* for the series, 'Handbooks (Textbooks) of Geography', which was edited by Erich Obst. However, he changed the title of his volume to *General Vegetation Geography*. Schmithüsen viewed vegetation geography as pertaining to ecosystems and types of vegetation as the aims of geographical research and representation. Schmithüsen's Handbook was published in three editions in 1959, 1961 and 1968, as well as in Japanese and Russian translations.

Schmithüsen was a pioneer in introducing vegetation geography as a main aspect of geographical research, vegetation now being represented as an important factor in the structure of landscapes. He also introduced methods from the fields of plant sociology and ecological site studies into the field of vegetation research. Schmithüsen's final and greatest accomplishment was the *Atlas of Biogeography*, published and illustrated by the Meyer Geographical-Cartographical Institute in 1976. In this atlas the potential vegetation constellations for all continents were mapped according to a uniform typology at a scale of 1: 25 million.

In Schmithüsen's many publications it is noteworthy how clearly he followed his own concepts and explained his thoughts on the subjects treated. In contrast to the many differing ideas relating to 'landscape' in everyday speech, Schmithüsen defined the geographer's 'landscape' as *the* qualitative unit of the geophysical landscape pattern. For Schmithüsen, the term 'landscape' is the basic scientific aspect of geography. The various opinions presented on the subject of geographical research were countered by Schmithüsen's requirement of a definitive clarification of the specific substance of the field of geography *per se*. The theme 'landscape' was extended to become a 'Manual of General Geography', or in the sense that Schmithüsen attached to the term 'study of landscapes', he termed 'General Geosynergetics', to avoid confusion with words used in everyday language. Schmithüsen termed 'landscape' (or 'geosynergose') as the aggregate of the conditions of a basic character of a part of the Earth or landscape region. This was defined by Alexander von Humboldt as the 'total character' of a physical region.

Schmithüsen's book *Manual of General Geography* represents a historical in-depth study of the development of teaching about 'landscape' as the character of an area of a specific landscape. The book also presents a multitude of new ideas and new definitions regarding geographical methods and aims.

In addition to the two manuals which Schmithüsen wrote, his final, crowning publication wherein he presents his ideas of the geographer's work, is the *International Geographical Glossary* (German edition published in 1985). Schmithüsen prepared the explanatory texts for 200 technical terms, carefully formulating the basic terms of landscape research, biogeography, and plant geography stemming from his life's-work. Schmithüsen did not live to see his work in print; he died of a heart attack on 2 September 1984, on returning from his second home in Formentera, Spain.

3. Influence and Spread of Ideas

Schmithüsen was the recipient of many honours and prizes for his outstanding work in the field of physical geography. He was awarded the Robert Lehmann Prize in 1952, during the seventieth jubilee of the Central Committee of German Geography, and the Robert Gradmann Medal in 1970 at the German Geographers' Congress in Mainz. On Schmithüsen's sixty-fifth birthday, a jubilee edition of his collected works, *Landscape and Vegetation*, was published by his colleagues at the Geographical Institute of Saarbrücken University.

Schmithüsen was a member of the Central Committee of German Geography from 1947 to 1970; from 1951 to 1970 he was a member of the Board of the Institute of Regional Geography. Schmithüsen was also a member of the UNESCO Programme 'Man and the Biosphere'; he received all of these appointments in recognition of his scientific achievements.

Bibliography and Sources

1. OBITUARIES AND ARTICLES ABOUT JOSEF SCHMITHÜSEN

Kroesch, Volkmar, 'Josef Schmithüsen: Geographer and university professor', *Jahrbuch Naturkunde*, Vol. 140 (1985), 253–6.

Meynen, Emil, 'Josef Schmithüsen on the occasion of his seventieth birthday', *Geogr. Z.*, Vol. 67 (1979), 2–11.

——, 'Josef Schmithüsen: World traveller and vegetation geographer', *Erdkd. Wiss.*, Vol. 79 (1986), 7–9.

Müller, Paul, 'Schmithüsen on the occasion of his sixty-fifth birthday', *Amazonica* (1974).

Philippi, G., 'Josef Schmithüsen', *Carolinea*, Vol. 44 (1986), 183–4.

Rathjens, Carl, and Müller, Paul, 'Early perceptions – The landscape as environment', *Biographica*, Vol. 16 (1979), 1–4.

——, 'Dear Josef Schmithüsen', *Biographica*, 16, 1–4.

2. *ARTICLES BY SCHMITHÜSEN ABOUT COLLEAGUES AND AUTOBIOGRAPHY*

'Michael Walter and his publications on the regional geography of southwestern Germany, In honour of his 75th birthday on 24 September 1951', Bericht zur deutschen Landeskunde, Vol. 10 (1951), 311–16.

'Leo Waibel', *Naturwissensch. Rundschau*, Vol. 5 (1952), 217.

'Leo Waibel - 22 February 1988 - 4 September 1951', *Die Erde. Zeitschrift der Gesellschaft für Erdkunde* (Berlin), 99–107.

'Franz Roth', *Nachrichtenblatt alter Münstereifler*, Vol. 33 (1958), 6–8.

'Mathias Schwickerath on the occasion of his seventieth birthday', *Natur und Landschaft*, Vol. 6 (1962), 13–14.

'Obituary of Gustav De Lattin; address given at the Biographical Colloquium on 1 September 1968', *Abhandl. der Arbeitsgemeinschaft der tier und pflanzengeogr. Heimatforschung*, 1 (1969), 3–5.

'Emil Meynen on the occasion of his eightieth birthday', *Berichte zur deutschen Landeskunde*. 56 (1982), 159–65.

(From home country to the world; review of studies, research and teaching.) (1981) 'Von Heimat und Welt. Rückblick und Studium, Forschung und Lehre. Vortrag gehalten am 30. Januar 1979, anläßlich des Festkolloquiums zu seinem 70. Geburtstag, Hrsg. von der Universität des Saarlandes' (1981), 21 pp.

3. *SELECTIVE BIBLIOGRAPHY OF WORKS BY JOSEF SCHMITHÜSEN*

(Articles which are marked with an asterisk are published in Schmithüsen, J., *Landschaft und Vegetation. Gesammelte Aufsätze von 1934 bis 1971*, Saarbrücken, 1974.)

a. *Regional Geography*

1938 'Introduction: A geographical overview of the Ahrweiler District', *Kunstdenkmäler der Rheinprovinz*, Vol. 17, 3–8.

1939 'Geographical introduction', itrd, *Deutsches Archiv für Landes- und Volksforschung*, Vol. 3, 1–3.

1940 *Luxembourg - Land, Natural Geography, National Groups and Rural Economy, Forschungen zur deutschen Landeskunde*, Vol. 34, 431 pp.

1955 'Geographical excursion through the District of Cochem', *Berichte zur deutschen Landeskunde*, Vol. 14, 119–78.

1958 'Luxembourg', *Großer Herder-Atlas*, 422–4.

b. *Biogeography: Vegetation and Plant Geography*

1934a *'Vegetation geography and pedology in southern France and eastern Spain', *Geogr. Zeitschrift*, Vol. 40, 409–22.

1934b 'Studies of vegetation in the woodlands of the western Rhenish Uplands', *Tharendter Forstliches Jahrbuch*, Vol. 34, 225–64.

1935 *'Plant sociological studies of sylviculture', *Der deutsche Forstwirt*, Vol. 17, 185–8.

1940 'Forests and settlements in prehistoric central Europe', *Zeitschrift für Erdkunde*, Vol. 8, 117–18.

1942 *'Plant research and ecological site investigation – importance for assessing the geography of cultural landscapes', *Zeitschrift der Gesellschaft für Erdkunde*, Berlin, Hr. 3/4, 113–57.

1948 'Plant damage in the drought year 1947; observations carried out in southern and western Germany', *Kosmos*, Hr. 4, 241–4.

1950a *'Some aspects of drought observations and effects on vegetation', *Mitteilungen der Floristisch-soziologischen Arbeitsgemeinschaft*, Hr. 2, 60–3.

1950b *'The climax problem viewed from the standpoint of landscape research; dedicated to Robert Gradmann on his eighty-fifth birthday', *Mitteilungen der Floristisch-soziologischen Arbeitsgemeinschaft*, NF2, 176–82.

1950c *'Plant sociology and plant phenology a hundred years ago: History of vegetation research in southern Germany', *Berichte zur deutschen Landeskunde*, Vol. 7, 370–4.

1950d 'A milestone in the history of forestry research in central Europe', *Berichte zur deutschen Landeskunde*, Vol. 8, 68–73.

1951a *'Present-day problems and practical responsibilities of research on vegetation sociology in Germany', *Berichte zur deutschen Landeskunde*, Vol. 9, 323–30.

1951b *'Germany's forested areas', *Geogr. Taschenbuch*, 217–28.

1953 *'Borders of Chilean vegetation regions', *Tagungsberichte und wissenschaftliche Abhandlungen Deutscher Geographentag*, 101–8.

1954a 'Forest formations of northern Chile', *Vegetativ, Tetic Geobotanica*, (V) VI, 479–86.

1954b 'Evergreen sclerophyllous plants in the subtropical winter rain areas of central Chile', *Rhododendron und immergrüne Laubgehölze Jahrbuch*, 23–42.

1955 'The flowering desert', *Universum*, 4/3, 62–6.

1956a 'Natural regions and landscapes observed from the viewpoint of the plant sociologist and geographer', *Berichte zur deutschen Landeskunde*, Vol. 16, 223–7.

1956b Areal Arrangement of Chilean Vegetation.

1957 *'Beginnings and aims of vegetation geography', *Mitteilungen der Floristisch-soziologischen Arbeitsgemeinschaft*, NF 6/7, 52–78.

1958 *'Problems of vegetation geography', *Tagungsbericht und wissenschaftliche Abhandlungen Deutscher Geographentag* (1957), 72–84.

1959 *General Plant Geography*, Lehrbuch der allgemeinen Geographie IV. 1961[2], 1968[3].

1960 *'Conifers in forest lands of the southern Andes', *Vegetation Acta Geobotanica*, Vol. IX, 313–27.

1965 'Forests of the cultural landscapes of the northern Palatine mountains', *Geographische Rundschau*, Vol. 17, 152–4.

1966 'Problem of vegetation history in Chile and New Zealand', *Vegetatio Acta Geobotanica*, Vol. XIII, 189–209. German edition: *Landschaft und Vegetation* (1974), 504–19.

1968 'Vegetation map of the Earth 1: 25 million', *Duden-Lexikon*, Vol. 8, 321–46.

1969a 'Vegetation (with a map of scale 1: 25 million)', *Meyers Kontinente und Meere. Band Mittel- und Südamerika*, 25–8.

1969b 'Vegetation (with a map of scale 1: 25 million)', *Meyers Kontinente und Meere. Band Sowjetunion*, 25–32.

1969c 'Overview of vegetation regions on the basis of observed forms and scientific analysis. Mathias Schwickerath's Methodological Studies', *Mitteilungen der Fränkischen Geographischen Gesellschaft*, 15/16, 494–7.

1970a 'Vegetation (with a map of scale 1: 25 million)', *Meyers Kontinente und Meere. Band Nordamerika*, 29–36.

1970b 'Vegetation (with a map of scale 1: 25 million)', *Meyers Kontinente und Meere. Band Mittel- u. Südamerika*, 25–8.

1970c 'Vegetation and landscape'. Lecture in the Series 'Our Earth', presented by the Austrian Broadcasting Company on 10 March 1970, *Vegetatio Acta Geobotanica*, Vol. XX, 210–13.

1970d 'Problems (determining) the genesis of South American biota'. Co-authored with P. Müller. Festschrift für Ernst Gentz. 109–22.

1972 'World geographical plant map', *Brockhaus Enzyklopädia*, Vol. XIV, 400–1.

1973a 'Interpretation of bioclimatic zonation by maps of vegetation types', Pre-Congress Indonesia Symposium: Planned Utilization of Lowland Tropical Forestry, 74–6.

1973b 'Vegetation (with a map of scale 1: 25 million)', *Meyers Kontinente und Meere. Band Australien, Asien, Inseln und Meere*, Vols 32–3, 386–90.

1975 'Purpose of worldwide vegetation mapping in the international research programme "Man and Biosphere (MAB)" ', Festschrift zum 70. Geburtstag von Erich Oberdorfer, 289–94.

1976 *Atlas of Biogeography*. Volume 3 of *Meyers großer physikalischer Atlas*. 80 pp.

1977a 'Vegetation maps – basic information sources in connection with worldwide problems of environmental damage', *Proceedings Int. Symposium in Tokyo on Protection of the Environment* (1974), 35–8.

1977b 'Speech for the defence in vindication of a Japanese tree', *Proceedings Int. Symposium in Tokyo on Protection of Environment* (1974), 483–5.

1977c 'Biogeological studies in the Bükk Mountains (Hungary)', Co-authors

J. Goergen and H. Schreiber. *Studica phytologica in honorem publicantis A. O. Horvat*, 43–52.

1978 'Competition as a limiting factor in remaining areas of old conifer stands, together with an overview of ecological consequences for forestry planning', *Erdwissenschaftliche Forschung*, Vol. XI, 124–34.

1979 'Expedition through prehistory and protohistory of biogeography', *Geogr. Taschenbuch*, 72–82.

1985a 134 explanatory texts for biogeography and vegetation geography, *Internationales geographisches Glossarium* (E. Meynen), Deutsche Ausgabe 1985. Italian edn 1988.

1985b 'Prehistory and protohistory of biogeography', *Biographien*, Vol. 20, 158 pp.

1986 'The natural physical world of central Europe', *Erdkd. Wiss.*, Vol. 79, 71 pp.

c. Natural Divisions (Determination of Physio-Geographical Regions)

1936 'Comments on the physio-geographical regions of the western Rhenish Uplands and adjoining areas', *Rheinische Vierteljahresblätter*, Vol. 6, 209–29.

1942 *'Landscape regions of Lothringen', *Deutsches Archiv für Landes- und Volksforschung*, Vol. 6, 64–73.

1944 *'Physio-geographical regions and their representation on 1: 200 000 maps of the topographical overview map series of the German Reich', *Berichte zur deutschen Landeskunde*, Vol. 3, 1–7.

1948a 'Fundamentals and rules for the analysis of physio-geographical regions of Germany and their representation at a scale of 1: 200 000', *Geogr. Landesaufnahme 1: 200 000 Richtlinien und Mitteilungen*, Vol. 1, 4.

1948b *' "Fliesengefüge" of "landscapes" and "ecotypes": Proposals on conceptual arrangement and nomenclature of landscape research', *Berichte zur deutschen Landeskunde*, Vol. 5, 74–83.

1949 *'Basic principles for analysis and representation of Germany's natural regions', *Berichte zur deutschen Landeskunde*, Vol. 6, 8–19.

1950 'Purpose and status of work on the natural regional divisions of Germany', *Sitzungsberichte der Akademie für Raumforschung und Landesplanung*, Vol. 1, 12–15.

1951 *Four-zone administrative map of Germany with physio-geographical regions, Status 1 December 1950, Scale 1: 1 Million (Co-author Emil Meynen).

1952a ' "Regions géographiques" and "physio-geographical regions" – methodological comparison', *Berichte zur deutschen Landeskunde*, Vol. 10, 1–6.

1952b Physio-geographical regions, Map 161, Karlsruhe; Geographical Survey.

1953 *Basic and Methodology: An Introduction. Handbuch der naturräumlichen*

Gliederung. Edited by E. Meynen and J. Schmithüsen. 1 S, 1–44.

1954 'Plant site-maps – basis of physio-geographical regions', *Umschaudienst*, Vol. 4, 62–6.

1961 Geographical Basis of the Historical Development of the Karlsruhe Region.

1965 Geographical Mapping 1: 200 000, Physio-Geographical Regions: Basic Considerations and Principles for Studies of Germany's Natural Regions and Their Cartographical Representation on Maps at a Scale of 1: 200 000.

1967a 'Reports on two symposia: International discussion regarding theoretical problems concerning landscape research in Slovakia', *Berichte zur deutschen Landeskunde*, Vol. 39, 122–4.

1967b 'Physio-geographical regions and areal landscape regions', *Berichte zur deutschen Landeskunde*, Vol. 39, 125–31.

1968 Map of the Physio-Geographical Regions of the Central Upper Rhine Valley.

1975 'Historical geography and "Physio-Geographical Regions" – applications to local history', *Oberrheinländische Studien*, Vol. III, 2–9.

d. Geography and Delimitation of Landscapes

1949 *'Landscapes in the logical system of geography' (Co-author: H. Bobek), *Erdkunde*, Vol. III, 112–20.

1951 'Areal organization of landscapes', *Geogr. Taschenbuch*, 370–1.

1954a 'Landscape and landscape area', *Handbuch der naturräumlichen Gliederung Deutschlands*, Vol. I, 2–5.

1954b *'Mental perceptions and cultural landscapes', *Berichte zur deutschen Landeskunde*, Vol. 12, 185–8.

1956 'Natural space as viewed by the plant sociologist and Geographer', *Berichte zur deutschen Landeskunde*, Vol. 16, 221–9.

1959 *'The system of scientific geography', Festschrift für Th. Kraus. 1–4.

1961a 'Importance of the term "economic forms", as Illustrated in new works on landscape research', *Berichte zur deutschen Landeskunde*, Vol. 28, 81–8; English trans.: *Universitus* (1968), 173–84.

1961b 'Nature and spirit of Landscapes', *Natur und Landschaft*, Vol. 5, 70–3.

1962 *'Suggestions regarding international geographical terminology based on geospheric synergism' (co-author E. Netzel), *Geogr. Taschenbuch*, 283–6.

1963 'The scientific term "landscape"', Festschrift für Otto Stocker, *Mitteilung der Floristisch-soziologischen Arbeitsgemeinschaft*, NF 10, 9–19; also in *Pflanzensoziologie und Landschaftsskizze und Landschaftsökologie*, Den Haag 1968, 23–43.

1964 'What is a "Landscape"?' Commencement lecture at the University of Saarbrücken. *Erdkd. Wiss.* Vol. 9, 24 pp; also *Saarbrückereden*, Vol. 10.

1966a 'Theoretical problems in geography', *Japanese Geography*, Tokyo.

1966b 'Geographical theories'. Round-table Conference, *Geographica Helvetica*, Vol. 21, 36–7.

1967 *Berichte zur deutschen Landeskunde*, Vol. 39, 125–31.

1968a 'The system of geography', *Universitas*, Vol. 10, 173–84.

1968b *'Determination of the essence of the term "landscape" as an object of research in the fields of biology and geography', *Archiv für Naturschutz und Landschaftsforschung*, Vol. 8, 101–12.

1970a 'Beginnings and aims of modern geographical science' (Commemorative volume for W. Grotelüschen), *Moderne Geographie in Forschung und Unterricht*, 9–20.

1970b 'History of geographical science from its beginnings to the end of the 18th century', *B.I. Hochschultaschenbücher*, 363/363a. 190 pp.

1970c 'The scope of geographical science' (Dedicated to Erich Otremba on the occasion of his 60th birthday), *Geographische Rundschau*, Vol. 22, 431–7.

1973a 'The quality of man's environment', *Kölner Geogr. Arbeiten*, Vol. 30, 65–71.

1973b 'Development of idea of landscapes in European painting; prehistory of the scientific terminology of landscape', *Geographie heute. Einheit und Vielfalt* (E. Meynen). Erdkd. Wiss., Vol. 33, 70–80.

1974 'What do we understand by the term "landscape ecology"?' *Tagungsbericht und wissenschaftliche Abhandlungen*, Deutscher Geographentag (1973), 409–12.

1975 'Perception of the synergose as a basic landscape element', *Geogr. Taschenbuch*, 247–8.

1976 *General Geo-Synergismus: Basis of the Study of Landscapes. Lehrbuch für Allgemeine Geographie*, Vol. XII. 349 pp.

1985 67 explanatory texts of basic geographical terms and landscape research. *Int. Geogr. Glossarium.* German edn 1985; Italian edn 19.

e. Economic Geography and Research on Cultural Landscapes

1934 'Forest groves of the western Rhenish Uplands: Remarks concerning the geography of the Rhenish cultural landscape', *Beiträge zur Geographie der Rheinlande*, z Rh, 4. 106 pp.

1936 'Distribution of agricultural and forested areas in the region from Zell on the Moselle River to the Hunsrück: With a map of scale 1: 25 000', *Atlas der deutschen Agrarlandschaft*, Hrsg. von E. Ostremba. 1962, sheet V 2 b.

1937 'Forest groves in the western Rhenish Uplands, and the possibility of their clearance', *Raumforschung und Raumordnung*, 10. S, 404–9.

1938 Development of Luxembourg's Agriculture, from the Geographical Point of View.

1939 'Differences in character in the physiognomies of cultural landscapes on the Walloon – German National Border', *Deutsches Archiv für Landes- und Volksforschung*, Vol. 2, 41–58.

1941 *'Life and work of the Moselle vine-growers', *Berichte zur deutschen Landeskunde*, Vol. 1, 21–6.

1943 'Suggestions for use of specific symbolic colours for mapping agricultural areas', *Berichte zur deutschen Landeskunde*, Vol. 3, 131–3.

1948 *'Effects of the drought summer of 1947 as an applied research aim', *Berichte zur deutschen Landeskunde*, Vol. 5, 37–52.

1950 'Susceptibilities of central European economic landscapes in the past and present', *Berichte zur deutschen Landeskunde*, Vol. 6, 56–8.

1953 'Mining and industry in the North Chilean desert', *Schemis*, Vol. 77, 3–8.

1954 *'Mental essence of cultural landscapes', *Berichte zur deutschen Landeskunde*, Vol. 12, 185–8.

1958 *'Remarks regarding problems of soil erosion in cultural landscapes', *Angewandte Pflanzensoziologie*, Vol. 15, 165–8.

1971 *'The concepts of formation and landscape in economic geography', *Heidelberger Geographische Arbeiten*, Vol. 36, 26–31.

1972 'The Geographical Institute of the Saarland and ecological research', *Umwelt Saar*, 14–18.

1975 'Inception of a practically applicable objective method for delimiting industrial landscape zones', Festschrift für E. Otremba. *Erdkd. Wiss.*, Vol. 41, 91–100.

1977 *Geography and Land Planning*, Schriftenreihe zur Orts- Regional- und Landesplanung, Vol. 29, Zürich.

1979 'Secular changes in the landscape caused by man and the results thereof – a geographical research field', Festkolloquium für Ernst Neef. *Trierer Geographische Studien*, Sonderheft 3.

Emil Meynen is a long-time editor of Orbis Geographicus.

Chronology

1909 Born in Aachen, 1 January

1914–27 Attended the gymnasium in Aachen

1927–39 Attended the Technical University in Aachen

1929 Member of the State Agency for Nature Conservation's excursion, 'Central Rhine Botanical Field Trip'

1929 Study at Bonn University; major in geography

1932 Dissertation: 'Woodlands of the Western Rheinish Uplands'

1932 Bonn University, Assistant/Senior Assistant, Institute of Geography

1933 First independent research field trip to southern France and eastern Spain

1939 Inaugural dissertation: Luxembourg – 'Land, Natural geography, National Groups and Rural Economy'

1939 Married Franziska Schmidt (died 1962), who bore him eight children

1941 Co-worker and assistant director, Department of Regional Geography, founded by Emil Meynen within the National Office of

	Cartography, Berlin; founder of the system mapping of 'Natural Regions', scale 1: 200 000
1942	Drafted into the German Army and sent to the Eastern Front
1943	Transferred to the Research Branch as scientific liaison officer
1945–46	Prisoner of war
1947	Lectureship at the Technical University, Karlsruhe
1947–70	Member, Central Committee of German Geography
1950	Paper on 'Nature and Spirit of the Landscape', presented at Symposium in Coburg
1952	First field excursions, studying vegetation geography in Chile and Brazil
1953–70	Member of the Board, Institute of Regional Geography in the Federal Institute of Research on Regional Geography and Regional Planning
1953	Associate professor of geography, Technical University Karlsruhe
1957–58	Second expedition to Chile, with Erich Oberdorfer and H. Kühlwein; return trip via Ecuador, Colombia and Venezuela
1959	Professor of geography, Technical University of Karlsruhe
1960	Excursion of four months' extent to Ireland, Norway, Sweden and Finland
1961–62	Visiting professor at University of Auckland, NZ
1962	Professor of geography, University of the Saarland at Saarbrücken, in alternation with Carl Rathjens, second chair in geography
1965	Research expedition of five months in Japan, together with Professor Reinhold Tüxen
1966	Married Jutta Witzmann
1966	Research expedition in eastern Australia, New Zealand and Japan
1967	Attended the International Symposium, 'Theoretical Problems of Biological Research' in Bratislava, Czechoslovakia, and the Symposium 'Physiogeographical Regionalization' in Moravany/ Piešt'any
1970	Excursion to Tunisia and stay in Bizerte
1971	Third excursion to Chile
1971	Excursion to Indonesia and southern India
1971	Study trip to the USSR, and attendance at a Symposium, Institute Professor Victor Sočova, in Irkutsk
1972	Foundation of the Biogeographical Institute at the University of Saarbrücken and initiation of the journal *BioGeographica*
1973	Visiting professor at the University of Toronto, Canada
1976	Attended Twenty-third International Geographical Congress in Moscow; subsequent expedition in Uzbekistan (Tashkent, Samarkand and Bukhara)
1977	Retired
1980	Research excursion in Brazil and Chile
1984	Died of heart attack at second residence in Formentera, Spain

Index

The index is divided into two parts:

1. A general index, including personal names, organizations, conferences, societies, and geographical concepts, theories and research.

2. A cumulative list of biobibliographies which includes all the geographers listed in volumes 1–14 inclusive.

1. GENERAL INDEX

2. CUMULATIVE LIST OF BIOBIBLIOGRAPHIES